CW01270369

Springer Series in Fashion Business

Series editor

Tsan-Ming Choi, Institute of Textiles and Clothing, Hong Kong Polytechnic University, Hung Hom, Kowloon, Hong Kong

This book series publishes monographs and edited volumes from leading scholars and established practitioners in the fashion business. Specific focus areas such as luxury fashion branding, fashion operations management, and fashion finance and economics, are covered in volumes published in the series. These perspectives of the fashion industry, one of the world's most important businesses, offer unique research contributions among business and economics researchers and practitioners. Given that the fashion industry has become global, highly dynamic, and green, the book series responds to calls for more in-depth research about it from commercial points of views, such as sourcing, manufacturing, and retailing. In addition, volumes published in Springer Series in Fashion Business explore deeply each part of the fashion industry's supply chain associated with the many other critical issues.

More information about this series at http://www.springer.com/series/15202

Chris K. Y. Lo · Jung Ha-Brookshire
Editors

Sustainability in Luxury Fashion Business

Editors
Chris K. Y. Lo
The Hong Kong Polytechnic University
Hung Hom
Hong Kong

Jung Ha-Brookshire
University of Missouri
Columbia, MO
USA

ISSN 2366-8776 ISSN 2366-8784 (electronic)
Springer Series in Fashion Business
ISBN 978-981-10-8877-3 ISBN 978-981-10-8878-0 (eBook)
https://doi.org/10.1007/978-981-10-8878-0

Library of Congress Control Number: 2018935855

© Springer Nature Singapore Pte Ltd. 2018
This work is subject to copyright. All rights are reserved by the Publisher, whether the whole or part of the material is concerned, specifically the rights of translation, reprinting, reuse of illustrations, recitation, broadcasting, reproduction on microfilms or in any other physical way, and transmission or information storage and retrieval, electronic adaptation, computer software, or by similar or dissimilar methodology now known or hereafter developed.
The use of general descriptive names, registered names, trademarks, service marks, etc. in this publication does not imply, even in the absence of a specific statement, that such names are exempt from the relevant protective laws and regulations and therefore free for general use.
The publisher, the authors and the editors are safe to assume that the advice and information in this book are believed to be true and accurate at the date of publication. Neither the publisher nor the authors or the editors give a warranty, express or implied, with respect to the material contained herein or for any errors or omissions that may have been made. The publisher remains neutral with regard to jurisdictional claims in published maps and institutional affiliations.

Printed on acid-free paper

This Springer imprint is published by the registered company Springer Nature Singapore Pte Ltd.
part of Springer Nature
The registered company address is: 152 Beach Road, #21-01/04 Gateway East, Singapore 189721, Singapore

Acknowledgements

We would like to thank Series Editor, Prof. Jason Choi (the Hong Kong Polytechnic University), invited us to contribute this edited book for the series of the fashion business. This is a precious opportunity to work with scholars, who are also sharing the same vision as us, to promote the sustainability practices and awareness for the fashion industry. We would like to thank all contributing scholars for their invaluable researches and thoughts. This book is partly supported by the Hong Kong Polytechnic University, funding G-UA5M.

Contents

Contributors

Lei Min Cai Jinling Institute of Technology, Nanjing, China

RayeCarol Cavender University of Kentucky, Lexington, USA

Cindy Cordoba Louisiana State University, Baton Rouge, LA, USA

Nigel P. Grigg Massey University, Palmerston North, New Zealand

Jung Ha-Brookshire Textile and Apparel Management, College of Human Environmental Sciences, University of Missouri, Columbia, MO, USA

Sheikh I. Ishrat Ara Institute of Canterbury, Christchurch, New Zealand

Nihal Jayamaha Massey University, Palmerston North, New Zealand

Joseph P. Jones School of Journalism, University of Missouri, Columbia, USA

Katie Baker Jones School of Design and Community Development, West Virginia University, Morgantown, WV, USA

Eunmi Lee Textile and Apparel Management, University of Missouri, Columbia, MO, USA

Stacy Hyun-Nam Lee Business Division, Institute of Textiles and Clothing, The Hong Kong Polytechnic University, Hung Hom, Kowloon, Hong Kong

Heejin Lim The University of Tennessee, Knoxville, TN, USA

Chris K. Y. Lo Business Division, Institute of Textiles and Clothing, The Hong Kong Polytechnic University, Hung Hom, Kowloon, Hong Kong

Chao Min School of Information Management, Nanjing University, Nanjing, Jiangsu, China

Roger Ng Fashion & Textile Technology—Clothing, The Hong Kong Polytechnic University, Kowloon, Hong Kong

Venkateswarlu Pulakanam University of Canterbury, Christchurch, New Zealand

Spencer S. C. Tao Institution of Textile and Clothing, The Hong Kong Polytechnic University, Kowloon, Hong Kong

Thomas C. C. Wong International Fur Federation, Kowloon, Hong Kong

Li Zhao Textile and Apparel Management, University of Missouri, Columbia, MO, USA

Yi Zhou Business Division, Institute of Textiles and Clothing, The Hong Kong Polytechnic University, Hung Hom, Kowloon, Hong Kong

Chapter 1
Opening: Sustainability and Luxury Brands

Chris K. Y. Lo and Jung Ha-Brookshire

Abstract This chapter highlights the importance of exploring sustainability issues for luxury brands, and then it summarizesthe key highlights of each chapter of this book (i.e., chapter 2 to 10). We start by addressing the question "Issustainability an attainable goal for luxury brands?". Through the discussion of the tensions between the luxurybrand traditional operations and sustainability, this chapter provides the reasons for luxury brand incorporatingsustainability goals into their operations, while continuing their focus on quality and artisanship.

Keywords Luxury brands · Sustainability · Summary

Is sustainability an attainable goal for luxury brands? The term luxury was originally associated with "a lifestyle of excess, indulgence and waste" (Dubois, Czellar, & Laurent, 2005; Hennings, Stam, & Wenting, 2013, p. 27; Kahn, 2009). At the same time, sustainability focuses on moderate or responsible consumption of our resources to ensure that our future generation would be able to meet their needs on the Earth. Therefore, "excess" or "wastes" certainly seems to have negative impact on what sustainability pursues. Luxury is also associated with rarity. Rare animals' skins or furs symbolize luxury; luxury products, in this case, require consuming rare animals. Because of these meanings of luxury, luxury is often symbolized social inequality and therefore sustainability is antithetical to or incompatible with the concept of luxury (Kapferer, 2010).

However, today's consumers demand for sustainable supply chain and luxury products consumers are not an exception. In fact, more recently, luxury brands were

C. K. Y. Lo (✉)
Business Division, Institute of Textiles and Clothing,
The Hong Kong Polytechnic University, Kowloon, Hong Kong
e-mail: tcclo@polyu.edu.hk

J. Ha-Brookshire
Textile and Apparel Management, College of Human Environmental Sciences,
University of Missouri, 137 Stanley Hall, Columbia, MO 65211, USA
e-mail: habrookshirej@missouri.edu

© Springer Nature Singapore Pte Ltd. 2018

C. K. Y. Lo and J. Ha-Brookshire (eds.), *Sustainability in Luxury Fashion Business*,
Springer Series in Fashion Business, https://doi.org/10.1007/978-981-10-8878-0_1

criticized for their unsustainable raw materials selection and use of endangered animal skins for apparel products. Consumers started boycotting such brands and it negatively affected luxury brands' images overall (Davies, Lee, & Ahonkhai, 2011). Following these events, luxury brands, such as Gucci, Saint Laurent, Prada, and Chanel, committed to develop comprehensive sustainability programs, including promoting the use of energy efficient light sources, adopting recycling practices in product supply chain, fur-free, and opting for environmental-friendly bioplastic ad native organic wool (Guercini & Ranfagni, 2013; Vij, 2016). A handful of luxury brands, such as Gucci, Stella McCartney, and Armani, are committed to go fur-free (Featherstone, 2017). These examples show that luxury brands must also pay attention to sustainability and create sustainable supply chain to meet the demands of today's consumers. As a matter of fact, a luxury market analyst reports that sustainability has become the focus and the biggest goal in the recent marketplace, hoping to improve their brands' reputation and products' values (Tutty, 2016).

While developing strategies for sustainable supply chain, luxury brands must be aware of a few critical differences in consumer expectations between luxury and mass market products. For example, luxury product consumers are thought not to believe recycle labeling is necessary because they believe that higher materials used in luxury products should guarantee the durability and a long life span (Catry, 2003, Beckham & Voyer, 2014; Kapferer & Bastien, 2015). In addition, luxury product consumers believed that recycled materials would reduce the values and quality of luxury products, which is quite different from mass product consumers (Achabou & Dekhill, 2013). Furthermore, consumers often are not concerned with sustainability when they purchase luxury products partly because they believe luxury products are already made sustainably (Achabou & Dekhili, 2013; Torelli, Monga, & Kaikati, 2012), and their perfectionism on product quality implies wastage of materials (Vigneron & Johnson, 1999).

With these challenges in mind, Fionda and Moore (2009) urged that luxury brands must focus on their own genetic materials to communicate sustainable nature of their products. That is, first, luxury brands are mostly being considered as heritage brands with long brand history and longevity of product life. For example, the fashion history of Hermes began in 1837, Louis Vuitton began in 1854, and Burberry's first store was opened in 1856 (Choi & Shen, 2017). Bourne (2010) believed that brand history becomes a significant market indicator for luxury brands, and it inspires consumers to engage in investment shopping. In fact, in the today's fast fashion marketplace, the luxury industry still achieves US $1 trillion sales, with 14% growth rate in 2015, and it is expected to have a 9% annual growth by 2020 (Bian & Co., 2015; D'Arpizio, Levato, Zito, & Montgolfier, 2015). Therefore, Fionda and Moore believe that luxury brands must make strategically long-term investments by incorporating sustainability goals to reduce long-term business risks.

Additionally, Fionda and Moore (2009) also points out that the original concept of luxury is exclusive, limited, and hence long lasting. The limited edition and extremely highly priced-products help luxury brands maintain exclusivity in the marketplace. Such exclusivity would then drive reduction in over-consumption of natural resources. Finally, the authors also show that given luxury brands have built a strong positive public relationship with customers via high-quality design, retailing, and service, luxury brands may continue to focus on quality and artisanship while incorporating sustainability goals.

With these thoughts in our mind, this book is comprised of nine unique research findings related to sustainability and luxury brands. Chapter 2 titled sufficient desire: the discourse of sustainable luxury is authored by Katie Baker Jones and Joseph P. Jones. This chapter brilliantly introduces the conflicting nature of sustainability and luxury after performing discourse analysis of (a) three luxury brands' [Eileen Fisher, Stella McCartney, and Brunello Cucinelli] advertisements, fashion shows, and online communications, and (b) the fashion media's presentations, such as magazines, videos, and blogs about the three luxury brands. The results were that there were three key themes in communicating their sustainability: (1) sustainability to communicate the ideological brand; (2) sustainability to communicate the ideological product; and (3) sustainability to hedge bets with low modality. Throughout these discourses, these three brands were emphasizing materiality, quality, the maker, closely managed supply chains, and place as the key intersections of sustainable luxury, and the authors conclude this chapter by stating both luxury and sustainability are "actively evolving discourses" and there is "discursive struggle where the definitions, limits, and forms of these concepts are constantly negotiated."

Chapter 3 is titled the marketing of sustainability and CSR initiatives by luxury brands: cultural indicators, call to action and framework and is authored by RayeCarol Cavender. In this chapter, Cavender proposes cultural indicators of sustainable luxury brand communication and sustainable luxury brand communication hierarchical framework after a case study of LVMH's marketing communication. Three key themes emerged as cultural indicators of LVMH's sustainable luxury brand communication: (a) emerging consumption paradigm; (b) shifting consumer values; and (c) educating consumers about sustainable luxury. With these cultural factors in mind, Cavender proposes the four-level hierarchical systems within sustainable luxury brand communication. The bottom or the first level is related to preservation and conservation, focusing on conserving rare resources and raw materials, as well as local heritage and culture. The second level focuses on process, progress and environmental impact, largely addressing supply chain transparency and traceability. The third level points out production location concerned with country of origin or cultural heritage. Finally, the highest level or the fourth level of communication focuses on partnership, collaboration, and innovation, highlighting their cause advocacy and achievements in sustainability. The author concludes the chapter with future research topics urging empirical support for such framework.

The Chap. 4 titled luxury fashion brands vs. mass fashion brands: data mining analysis of social media responses toward corporate sustainability shows how luxury fashion brands communicate their corporate sustainability activities in the social media realm in general. Stacy Lee, Yi Zhou, Chris K.Y. Lo, and Jung

Ha-Brookshire conducted data mining analysis of Twitter messages posted by 11 fashion luxury brands, including Armani, Burberry, Cartier, and Chanel, and 11 mass fashion/sportswear brands, including Adidas, Gap, H&M, and Levis. After analyzing over 89,000 tweets and retweets under sustainability-related keywords, the authors found that mass fashion brands were more active on Twitter than luxury fashion brands in general, although luxury brands' sustainability messages were more often liked or retweeted. More interestingly, luxury brands were more associated with terms, donate, and charity, while mass brands were with recycle, fair, or planet. These findings are consistent with Cavender's [Chap. 3] finding of consumers' lack of knowledge about sustainability related to luxury products.

Next five chapters discuss more specific sector within the luxury industry. More specifically, Chap. 5 shows consumers' responses to animal cruelty in a luxury fashion supply chain, and Chap. 6 assessed consumers' social media responses to animal cruelty issues during the cosmetic product development processes. Chapter 7 deals with value chains of the cashmere industry, and Chap. 8 sheds lights into sustainability issues within the fur industry. Finally, Chap. 9 illustrates the drivers and barriers of luxury sector, specifically by looking at luxury retailers' the energy efficiency technology adoption behavior, and Chap. 10 describes artisans in Colombia who communicates not only luxury products but also social meaning of artisans' creativity in a local-specific context.

Chapter 5 titled luxury cruelty: an analysis of consumers' responses to animal cruelty in a luxury fashion supply chain is authored by Heejin Lim. Within the attribution theory framework, Lim analyzed consumers' responses to animal cruelty within the luxury fashion supply chain by coding and synthesizing a PETA's [People of Ethical Treatment of Animals] YouTube video which received more than one million views and 322 comments. This video shows the process of the killing of baby ostriches in a slaughterhouse for the ostrich skins that are used for luxury brands, such as Prada and Hermès. The results clearly showed that consumers were experiencing conflicting values [as Jones and Jones suggested in Chap. 2] between hedonic, materialistic, and conspicuous values of luxury and altruism, abstinence, and moral values of sustainability. Lack of awareness of unsustainable supply chain was highlighted once again, and therefore, the author concludes this chapter with strong needs of luxury brands' ethical material sourcing and transparency.

Chapter 6 titled mining social media data to discover topics of sustainability: The case of luxury cosmetic brands and animal testing is authored by Min Chao, Eunmi Lee, and Li Zhao who reviewed Instagram and Twitter messages posted under the #animal testing with the cosmetic product development contexts. After social network analysis, the authors found that Instagram users discussed animal testing are associated with terms, such as #Pestagram, #Save a life, #Vegan, #Cruelty-free, #Ethical. The results of the Twitter data analysis were consistent with the Instagram messages; however, 3D printing or biotechnology was discussed as alternative options for animal testing. Given that the Instagram and Twitter have different sets of user profiles, the authors explained different results expected.

Chapter 7 titled cashmere industry: Value chains and sustainability, discuss how to improve cashmere manufacturing processes while integrating sustainability goals. Sheikh Ishrat, Nigel Grigg, Nihal Jayamaha, and Venkateswarlu Pulakanam reviewed the value chains of cashmere products in both traditional and modern practices. The authors show that automation of fiber processing, transformation, labeling, and recycling stages negatively impacted not only the quality of the products due to unscrupulous manufacturing processes, but also social imbalance due to cashmere goat herders' abandonment of traditional cashmere production and migration into urban centers. To enhance the overall sustainability within the cashmere value chains, the authors propose new research topics for the future researchers: How can we balance modernization of the value chain while protecting and preserving environmental resources and social community?; how can we balance the degree of fiber blends to increase demand while protecting unique values of cashmere fibers?; and how can we balance the impact of globalization and localization of the cashmere value chains?

Chapter 8 is titled sustainability in the fur industry and is authored by Thomas Wong, Roger Ng, and Lei Min Cai. The authors interviewed four key leaders of the fur industry and executives of fur companies for their views on the fur industry from the sustainability perspective. The results showed that these executives shared that fur products could be sustainable and the fur industry could be sustainable if certain efforts are made. For example, one of the key topics raised from the interviews was that there is a need for sustainable demand for products as luxury products. With the prevalence of faux-furs in the marketplace, the participants agree that fur must be communicated as heritage products that have high-performance function. Manufacturing process must be sustainable as well and increased transparency will help enhance the sustainability nature of the fur industry. Fur could also be communicated as an effort to preserve local communities as it was warned by the authors of Chap. 7. In this light, the authors conclude this chapter with a proposal of a holistic framework of the fur industry's sustainability efforts where both economic and political environments in the global marketplace could have more harmonious relationships with the fur supply chains.

Chapter 9 showcases the drivers and barriers of luxury retailers' adoption of energy efficiency technology. The chapter is titled the drivers and barriers of luxury sector retailers to adopt energy efficiency technologies in Hong Kong. Spencer S.C. Tao and Chris K.Y. Lo interviewed six senior executives of luxury retailers in Hong Kong and explored their energy efficiency technology adoptions. Some of the key drivers of such adoption were corporate social responsibility, top management's commitment, corporate mission as a global company, and responses to external incentives. The barriers to such adoption were lack of control over the building, competitive market conditions, low-cost savings, and lack of awareness. Based on the institutional theory, the authors propose a new conceptual model of energy sustainability strategy for luxury retailers, which predicts that normative, coercive, and mimetic pressure would influence luxury retailers' energy efficiency technology adoptions, which in turn, would result in lower carbon footprint and higher profitability. The authors encourage future research that would test these relationships.

Finally, Chap. 10 is titled luxury fashion business and the aim of peace restoration for artisans in Colombia and is written by Cindy Cordoba Arroyo. The author reviewed "Maestros Costureros" [or Master Sewers] program which was designed to bring artisans and designers to collaborate and co-create luxury products that symbolize a commitment to cultural heritage, peace, and economic development in Colombia. Through this case study, this chapter shows the interconnection between artisanship and fashion design as a way of increasing revitalization of artisan heritage techniques [directly related to slow fashion movement] and economic development efforts via luxury products. Therefore, the author describes that, with together, more sustainable fashion products are being produced with multiple-stakeholders' commitment and participation. The author concludes the chapter with an additional value that luxury fashion products produced from this program: that is, Master Sewers represents a peace-building strategy that elevates cultural value of artisans who were negatively affected by the Colombian conflicts, suggesting perhaps luxury products could sometime mean more than products themselves.

With these chapters, this book intends to highlight the challenges and opportunities that luxury brands and luxury products may have with regard to sustainability. Some show inherent conflicts in values between luxury and sustainability. Others point out specific value chain drivers and barriers of a variety of luxury products. We hope that the readers will be able to gain comprehensive understanding of sustainability and luxury and hope the readers will be inspired to have more discussion about the two and this book would be a catalyst to future role of the luxury industry in our goals toward sustainability.

References

Achabou, M. A., & Dekhili, S. (2013). Luxury and sustainable development: Is there a match? *Journal of Business Research, 66*(10), 1896–1903.

Beckham, D., & Voyer, B. G. (2014). Can sustainability be luxurious? A mixed-method investigation of implicit and explicit attitudes towards sustainable luxury consumption. *NA—Advances in Consumer Research, 42,* 245–250.

Bian & Co. (2015, October 29). Currency fluctuations and luxury globe-trotters boost global personal luxury goods to over a quarter trillion. Retrieved from http://www.bain.com/about/press/press-releases/Luxury-Report-Fall-2015-Press-Release.aspx.

Bourne, L. (2010, August 10). Fashion history with a twist: The world's most en during brands. Retrieved from https://www.forbes.com/2010/08/04/louis-vuitton-logo-hermes-lbirkin-ferragamo-shoes-burberry-trench-forbes-woman-style-luxury-fashion-brands.html.

Catry, B. (2003). The great pretenders: The magic of luxury goods. *Business Strategy Review, 14* (3), 11–17.

Choi, T. M., & Shen, B. (2017). *Luxury fashion retail management.* New York, NY: Springer.

D'Arpizio, C., Levato, F., Zito, D., & Montgolfier, J. (2015, December 21). Luxury goods worldwide market study fall-winter 2015: A time to act? How luxury brands can rebuild to win. Retrieved from http://www.bain.com/publications/articles/luxury-goods-worldwide-market-study-winter-2015.aspx.

Davies, I. A., Lee, Z., & Ahonkhai, I. (2011). Do consumers care about ethical-luxury? *Journal of Business Ethics, 106*(1), 37–51.

Dubois, B., Czellar, S., & Laurent, G. (2005). Consumer segments based on attitudes toward luxury. *Marketing Letters, 16*(2), 115–128.

Featherston, E. (2017). Gucci commits to going fur-free from 2018. Retrieved from http://www.independent.co.uk/news/business/news/gucci-fur-free-2018-commitment-social-responsibility-italian-fashion-brand-marco-bizzarri-a7996061.html.

Fionda, A. M., & Moore, C. M. (2009). The anatomy of the luxury fashion brand. *Journal of Brand Management, 16*(5–6), 347–363.

Guercini, S., & Ranfagni, S. (2013). Sustainability and luxury: The Italian case of a supply chain based on native wools. *Journal of Corporate Citizenship, 2013*(52), 76–89.

Henning, M., Stam, E., & Wenting, R. (2013). *Path dependence research in regional economic development: Cacophony or knowledge accumulation?* (Papers in Evolutionary Economic Geography 12, No. 19). Utrecht: Utrecht University Urban & Regional Research Centre.

Kahn, J. (2009). *Luxury-goods makers embrace sustainability*. New York Times, 26.

Kapferer, J. N. (2010). *Luxe et développement durable peuvent-ils faire bon mé nage? La Tribune*. May 2010.

Kapferer, J.-N., & Bastien, V. (2015). *The luxury strategy: Break the rules of marketing to build luxury brands* (2nd ed.). London: Kogan Page.

Torelli, C. J., Monga, A. B., & Kaikati, A. M. (2012). Doing poorly by doing good: Corporate social responsibility and brand concepts. *Journal of Consumer Research, 38*(5), 948–963.

Tutty, J. (2016, January 20). 2016 Predictions for the luxury industry: Sustainability and innovation Positive Luxury. Retrieved from http://blog.positiveluxury.com/2016/01/20/2016-predictions-luxury-world-sustainability-innovation/.

Vigneron, F., & Johnson, L. W. (1999). A review and a conceptual framework of prestige-seeking consumer behavior. *Academy of Marketing Science Review, 1999*, 1.

Vji, V. (2016, February 19). Social responsibility the knowledge of luxury [Web log post]. Retrieved from http://luxuryinstitute.com/blog/?cat=338.

Chapter 2
Sufficient Desire: The Discourse of Sustainable Luxury

Katie Baker Jones and Joseph P. Jones

Abstract Luxury and sustainable fashion goods are like any consumer product; they are material culture, laden with meaning beyond their utility. The value of a fashion good—whether luxurious, sustainable, or both—is largely symbolic rather than economic (Crane & Bovone in Poetics 34:319–333, 2006), though the economic value allows for a greater distinction between levels of goods. Consumers utilize these embedded values to make their personal values visible to themselves and others. Utilizing discourse analysis, we offer here an explication of the separate yet interlacing domains of 'luxury' and 'sustainability.' This is an attempt to locate, if only briefly, the shifting domains of these oft-employed concepts. Stella McCartney, Eileen Fisher, and Brunello Cucinelli served as case studies for this analysis due to their general ascension as preeminent sustainable luxury fashion brands. Central to this success has been their ability to use discursive practices that effectively communicate their company ethos. Whether that ethos is then connected to their product is questionable and will be explored here.

Keywords Sustainable luxury · Discourse analysis · Sustainable fashion

Introduction

Luxury and sustainable fashion goods are like any consumer product; they are material culture, laden with meaning beyond their utility. The value of a fashion good—whether luxurious, sustainable, or both—is largely symbolic rather than economic (Crane & Bovone, 2006), though the economic value allows for a greater

K. B. Jones (✉)
School of Design and Community Development, West Virginia University,
704 Allen Hall, Morgantown, WV 26506-6124, USA
e-mail: kathryn.jones@mail.wvu.edu

J. P. Jones
School of Journalism, University of Missouri, Columbia, USA

© Springer Nature Singapore Pte Ltd. 2018

C. K. Y. Lo and J. Ha-Brookshire (eds.), *Sustainability in Luxury Fashion Business*,
Springer Series in Fashion Business, https://doi.org/10.1007/978-981-10-8878-0_2

distinction between levels of goods. Consumers utilize these embedded values to make their personal values visible to themselves and others.

As ideological concepts, "sustainability" and "luxury" are not natural bedfellows. They have divergent definitions, applications, and discursive practices. However, within the domain of consumer goods—particularly fashionable goods—these discourses have increasingly merged, morphing the meaning of both. Utilizing discourse analysis, we offer here an explication of the separate yet interlacing domains of "luxury" and "sustainability." This is an attempt to locate, if only briefly, the shifting domains of these oft-employed concepts. Stella McCartney, Eileen Fisher, and Brunello Cucinelli served as case studies for this analysis due to their general ascension as preeminent sustainable luxury fashion brands. Central to this success has been their ability to use discursive practices that effectively communicate their company ethos. Whether that ethos is then connected to their product is questionable and will be explored here.

Meaning production through discourse is a complex social process. Jorgensen and Phillips (2002) overview of post-structuralist discourse theory offered a succinct explanation of how cultures come to know the world around them through discourse. With each new utilization, there is an attempt to (re)establish concepts in a web-like relationship to other concepts. However, vague concepts are more likely to morph from discourse to discourse. In fact, it is common for competing discourses to emerge when comparing across sites or applications (Macdonell, 1986). Both luxury and sustainability are fuzzy-edged terms that in many ways contradict one another.

Regardless, they have moved increasingly into closer proximity through fashion discourse creating "sustainable luxury." We explored this process in situ via case studies. Sites of analysis included consumer-facing narratives on Web pages (i.e., "About Us" pages and product descriptions) as well as external discourses profiling the brand creators located in major media outlets like The New Yorker and The New York Times Magazine. However, we first outline our discourse analysis approach since the literature review is itself a discourse analysis and should be understood as such.

Literature Review

Discourse Theory and Analysis

"Any discourse concerns itself with certain objects and puts forward certain concepts at the expense of others" (Macdonell, 1986, p. 3). This epigrammatic statement offered the most direct line to understanding discourse analysis as a method. Foucault argued that it is through discourse and discursive practice that we know what we know and what we know then shapes discourse (Foucault, 1970). Thus, it would serve to reason, to understand the reality of our own making, one must look to discourse. Additionally, post-structuralist discourse analysis incorporates intertextuality, or "the way that the meanings of any one discursive image or text depend not only on that one text or image, but also on the meanings carried by other images and texts" (Rose, 2001, p. 136). The discursively constructed domains of luxury and sustainability as they relate to fashionable goods, specifically apparel and accessories, are particularly prime for analysis because of each concept's potentials, limitations, and meanings. Thus, this research is an explication of those terms within the specific domain of fashion. We will analyze how brand communications via product pages and corporate philosophy, as well as external fashion media such as magazine editorials, construct, shape, and appropriate the concepts of sustainability and/or luxury for three "luxury sustainable fashion" brands.

There are multiple presuppositions when engaging in discourse analysis as outlined by and Jorgensen and Phillips (2002). First, discourse is naturally social and thus is co-constructed between participants. Second, its position as a co-construction between two or more parties—whether institutions like fashion brands or individuals like consumers and fashion journalists—means language is not a reflection of a preexisting reality, but a construction of it. Third, while discourse has structure and patterns, it is through discursive practices that patterns are maintained or transformed. Thus, discourse analysts interpret patterns and observe how they shift, morph, or contradict within and across contexts. Freeman, Martin, and Parmar (2007) noted in discourses on capitalism that "… the way we talk about markets and the assumptions we make about value creation also play a role in creating the outcomes we want and also those we do not" (p. 304). While discourse analysts engage with discourse in context, they also create context (Jorgensen & Phillips, 2002), serving as both frame and content. Thus, discourse analysts should maintain a constant position of reflexivity to account for their role in constructing new discourses through research (Fairclough, 2001).

Fashion discourse exists at the intersection of production and consumption. "Text contributes to an understanding of fashion by assigning descriptive or interpretive meanings to the objects and images presented on fashion pages, thereby mediating a cultural understanding of the phenomenon" (König, 2006, p. 207). As fashion information dissemination has developed through a growing media, the diffusion cycles have sped up immensely with increased complexity (Kawamura, 2005).

There are two main institutional sources of consumer-facing fashion discourse: (a) the designer or brand's advertising, fashion shows, and corporate (online) communication and (b) the fashion media's various outlets such as magazines, videos, and blogs (Kawamura, 2005). We examined three avenues of discourse construction for each brand: (a) the "About Us" or equivalently titled sections on each brand's corporate homepage; (b) five online product descriptions each for five different product categories (sweaters, denim, tops, outerwear, and shoes) across the three brands; and (c) three magazine editorials profiling each namesake designer (Horyn, 2012; Malcolm, 2013; Mead, 2010).

We examined the texts included in this explication both syntagmatically and paradigmatically (Fairclough, 2001, p. 240). We attempt to move behind the surface to assess the ideologies, values, and power structures informing the discourse. We assessed the reality constructed, the dynamic created, and ideological work (Fairclough, 2001) achieved through the discourse. As Chaffee (1991) noted, "Explication should tell us, among other things, the extent to which we are falling short of studying what we really intend" (p. 5). In many ways, luxury and sustainability are "primitive terms," commonly understood or taken for granted; "they are not questioned within the framework of research built upon them" (Chaffee, 1991, p. 7). Through the discussion herein, we hope to undermine the "naturalness" of the sustainability/luxury pairing. We do not intend to offer a solution to the contradiction contained within their association but rather call attention to the ideological work behind their coupling. Furthermore, we question that which is gained and/or lost in each concept's meaning when their discourses are intentionally combined to ideologically enhance a brand's offerings.

The Lexicon of Luxury

Luxury is a term taken for granted at the individual level. As an empty signifier (Laclau, 2006), it easily morphs from context to context. The *New Oxford American Dictionary* provided the following definitions: (1) "the state of great comfort and extravagant living" or (2) "an inessential, desirable item that is expensive or difficult to obtain" (Luxury, 2010). Berry (1994) offered the broadest perspective of luxury as concept in a historical and contextual overview of its cultural evolution. He argued that goods deemed luxurious stemmed from one of the four basic need groups: food, shelter, clothing, and leisure. Luxury builds upon basic needs in socioeconomically stratified societies, offering an outlet for differentiation (Veblen, 2004). While one may need to eat to live, eating a dry-aged porterhouse steak to satiate hunger is a luxury. Because the exploited need is basic or essential, brands focus the luxury discourse on the desirability of one good over another.

Academically, a consensus has formed around the characteristics necessary for a consumer to deem a good luxurious. As Godey et al. stated, "The common denominators are beauty, rarity, quality, and price, and also an inspirational brand endorsing the product" (2012, p. 1462). Again, each of these individual concepts is vague, relative, and subjective, doing little to solidify the concept at hand. For example, at Hermès rarity is as much a cultivatable trait through discursive practice as it is a production reality. By "frustrating demand" for its ever-popular handbags, Hermès created an illusion of rarity across all product categories including readily available and relatively easy-to-produce items like scarves (Solca, 2017, February 14).

Thorstein Veblen (2004) postulated in the early twentieth century that consumers acquire luxury goods as a mechanism to communicate social status to their peers through the process of conspicuous consumption. However, economists Bagwell and Bernheim (1996) claimed there is no intrinsic difference between mass and luxury goods indicating other processes must account for monetary differences. The fashion industry has relied on media to instruct, educate, and entice its potential consumers for centuries. The modern incarnation of glossy pages—whether a tangible magazine or corporate Web site—combines overt advertising, descriptive text, and seductive photograph spreads to define a brand and its products as luxury.

As a marketing keyword, luxury has enveloped a wide array of goods, sending an invitation to individuals outside the traditional target socioeconomic sphere to participate in this discourse. In their work on "New Luxury," Silverstein Fiske and Butman (2003) noted as the middle market grew, new opportunities arose to parlay concepts of luxury into the mass market. Geared toward the middle market, those who could not afford the very best in every category could splurge on "affordable luxury" in the categories that mattered to them. This "New Luxury" exemplified the subjective nature of luxury as it operated by mechanisms of pleasure and desire. This contrasted with "Old Luxury" "based primarily on status, class and exclusivity rather than on genuine, personal emotional attachment" (p. 7). Thus, through discursive practice, a new definition of luxury emerged that challenged the old concept, yet maintained its basic logic to engage a larger audience.

With New Luxury on the rise, it is even more imperative for Old Luxury brands to establish distinct differences between their offerings and those of emerging brands attempting to emulate their prestige. Discourse in luxury fashion magazines, which arguably subscribe to the tenets of Old Luxury, consistently emphasize time as a key characteristic of luxurious goods and thus central to their desirability. That is, one must wait, sometimes for months, before satisfying the desire for a truly luxurious good. Designer ready-to-wear (RTW) shows occur six months before delivery to stores. Though couture shows occur closer to the intended season, clients must still wait for their selected goods to be handcrafted and forfeit the pleasures of instant gratification. New Luxury, on the other hand, must be delivered in the timely manner demanded by its mass-market consumers. The innovation of "see-now-buy-now" runway shows was the ultimate expression of this drive. European brands were concerned "see-now-buy-now" reduced the "specialness" of their luxury offering while American RTW welcomed the innovation (Lockwood, 2016, p. 10, October 12). With the prevalence of fast fashion and the increasing pressure on designers to produce more collections per year, Old Luxury need only highlights the time investment in their goods to offer a point of difference.

While luxury as concept has morphed from Old to New, the physical manifestation of luxury lost ground is in the realm of artisanship and materiality (i.e., process and substance). As Thomas (2007) noted, since the 1960s luxury brands have used a variety of techniques to increase profits and democratize their offerings. This included outsourcing production to factories in developing countries, using lower quality or non-luxurious materials, diversifying product lines to include lower-priced aspirational purchases (i.e., cosmetics and fragrance) and spending more on advertising their brand image.

It is within this diminished sector of process and substance where one can envision the usefulness of coupling sustainability and luxury. Previous scholars have postulated they are not in conflict and can greatly benefit from each other. Cherny-Scanlon (2016) argued that the two domains already shared many of the same values: "respect for tradition and craftsmanship, the preference given to quality over quantity and the quest for harmony between humans and nature" (p. 184). She does not, however, offer any evidence of this last point. Neither does she acknowledge the lack of transparency in the luxury fashion sector (Fashion Revolution CIC, 2017) which does little to quell doubts as to the breadth and depth of any one brand's sustainability initiatives.

Sustainability as a Discursive Domain

Sustainability, like luxury, is a word used frequently and across a diverse array of fields. The OED lists three definitions for "sustainable/sustainability." The third, which is most relevant to this context, has both a generic and a human-centric orientation. First, the OED defines sustainability (adj.) as an act that is "capable of being maintained or continued at a certain rate or level." The more human-centric definition states that sustainability "designates human activity (especially of an economic nature) in which environmental degradation is minimized, especially by avoiding the long-term depletion of natural resources, of or relating to activity of this type."

Human-centric definitions informed several of the specific applications of sustainability in various contexts. Today, the most oft-cited definition of sustainability originated in the "Brundtland Report." This report, however, did not define sustainability but instead addressed sustainable development that "meets the needs of the present without compromising the ability of future generations to meet their own needs" (World Commission on Environmental Development (WCED), 1987). The report emphasized the need for economic growth but suggested setting limits and taking advantage of technological innovation to ensure damage minimization.

The human-centric definition of sustainability is problematic for three reasons. First, it assumed primacy of the human species in the hierarchy of the natural world and that preservation of human systems is primary. Second, it presupposed a universality of human needs and on the type of 'development' needed. Third, it assumed economic development was still the primary concern. This classifies the human-centric definition of sustainability as "weaker sustainability" (Williams & Millington, 2004, p. 101). As Williams and Millington explained, though all definitions of sustainable development concern the "environmental paradox"—the difference between demands on the Earth and what the Earth can provide—solving the paradox usually took one of two routes (p. 100). The "Brundtland Report" represented the first route: efficient resource use and relying on technology to meet demand. There is little concern for curbing demand, merely finding better ways to meet it. "Stronger sustainability," however, required reduced demand and divesting of the human-centric perspective (p. 102).

"Sufficiency" emerged as a helpful concept to operationalize "stronger" sustainability measures. In The Logic of Sufficiency, Princen (2005) summarized this perspective:

> This is the need for a language consonant with 'enoughness' and 'too muchness,' not just words, but concepts and organizing principles. In an ecologically constrained world, people need the rhetorical and political means for turning a silencing hand to the barkers and boosters, to the marketeers, to the spinmeisters and political handlers, all of whom tell us that the good life comes from purchasing goods, and that because goods are good more goods must be better" (p. 6).

Bocken and Short (2016) further explored sufficiency in the contemporary market and argued it is a viable business strategy with a loyal consumer base willing to pay a premium for fewer, more durable goods. This mentality harkens back to the Old Luxury model, when resources were scarce and well-trained artisans were central to production.

Luxury companies have succumbed to the fast-fashion model as much as any fashion price point. The system efficiently moves raw materials and reduces the cost of labor inputs. However, its efficiency has led to incredibly cheap clothing and an endless design cycle resulting in "fast fashion" (Clark, 2008). Consumers now tend to buy on sale, choose quantity over quality, and use/discard without thought to the impact on social and environmental systems. Luxury brands have opened outlet stores, paired with fast-fashion companies, and increased the number of seasons from two to as many as six a year. This perpetual cycle has also led to designer burnout (Cochrane, 2015, November 1; Heiderstadt, 2015, October 28), a different though equally important type of "waste," that of creative talent. Constant change is the hallmark of luxury fashion and is central to its unsustainable nature.

To find solutions to the rampant waste in the industry, scholars have explored the designer's role conceiving more environmentally friendly products. Designers, according to Walker (2010), should rethink their production process so they can either quickly adapt to change or have their products easily broken down for reuse and recycling. While this attention to production is a workable solution within the current paradigm, this strategy does little to curb demand for products and the use of natural resources.

Similar to the ecological issues facing the fashion industry, awareness of worker exploitation has also increased over time (Dybicz, 2004). With each new revelation —such as the recent tragedies in the Bangladesh textile industry in 2013 (de Graaf, Wann, & Naylor, 2014, p. 69)—there is a surge in public outrage over the working conditions used to make fashion goods.

"Social responsibility" as a discourse emerged from business literature and is frequently referred to as Corporate Social Responsibility (CSR). CSR is as contested a concept as sustainable development and sustainability (Dobers & Springett, 2010). CSR has led to changes in the business lexicon such as referring to "stakeholders" rather than "shareholders." This discursive change challenged the dominant capitalist discourse in regard to who should benefit from market operations (Freeman et al., 2007). However, "[t]he blurring between social responsibility and profitability constrains the level of commitment to change that companies can make, the emphasis generally being on picking the low-hanging fruit of cost-savings, efficiencies and PR that pose little threat to the status quo" (Dobers & Springett, 2010, p. 65). Dobers and Springett (2010) cautioned against being "distracted" by CSR discourse that denotes corporate initiatives "laudable in their own right but which do not make a difference to 'the way things are'" (p. 65). Dobers and Springett's perspective instructs the critical approach underlying this research, as well. While any action taken by a luxury fashion goods manufacturer to improve their processes in terms of environmental or social impact is commendable, we argue those actions should also be critically evaluated for how disruptive or compliant they are to traditional, unsustainable processes.

Sustainable, or "slow," fashion emerged in both academic and fashion discourse as a potential resolution to the waste and insatiability of the current system. Hazel Clark (2008) argued that if societies emphasized a personal connection to material culture, a slower fashion process would emerge. Clark reasoned, "the emotional attachment between human beings and clothes offers potential for designers wanting to explore fashion as a sustainable practice" (p. 441). She argued material goods must become more important, not less, to alter the way individuals and societies view the clothing they consume.

Slow consumption emerged as an ethos for those promoting a different relationship between consumers and their goods. Temporality is central to the slow consumption movement offering a potential link to aspects of luxury discourse. Slow consumption "means slowing the rate at which products are consumed (literally 'used up') by increasing their intrinsic durability and providing careful maintenance" (Cooper, 2005, p. 54). Slow consumption in fashion includes wardrobe building rather than trend following, choosing laundering options that preserve the material longer, and repairing the mendable. These practices require an educated consumer who is knowledgeable of the materials and processes used to create and care for fashionable goods. As noted above, media and brand discourse is one potential source for this knowledge.

Below, we help reveal what type of knowledge is gained through sustainable luxury brand discursive practices. Furthermore, we explore how the seemingly incongruous discourses of luxury and sustainability are merged, accommodating some aspects of each discourse while ignoring others. We offer a visual representation of the merged domains after exploring the specific discursive practices employed by three fashion brands currently operating within the sustainable luxury paradigm.

Case Studies

From Ideology to Object: Discursive Practices for Three Sustainable Luxury Fashion Brands

Stella McCartney, Eileen Fisher, and Brunello Cucinelli are known in both industry and the academy as companies operating within/between the realms of sustainability and luxury (Bocken & Short, 2016; Curwen, Park, & Sarkar, 2012; Kapferer

& Michaut, 2015). These companies intrinsically link the companies' initiatives and their founders' ideologies. Thus, they are sites in which the luxury/sustainability discourse emerged from within, rather than from external market or consumer forces. This internal drive aligns them with the core strategy for effective sufficiency initiatives as proposed by Bocken and Short (2016). We offer representative examples of the discourse created by, for, and around these fashion brands to illustrate the discursive practices employed to build a sustainable luxury fashion narrative which ideologically infuses their products with socially valued meanings. We examine (a) the discursive processes that connect an ideology to each luxury brand; (b) how each luxury brand's products are infused with ideology; and (c) specific discursive strategies employed by each brand to frame their sustainability and luxury narratives.

The ideological brand

Through corporate discourse all three brands constructed a narrative of "stakeholder capitalism" (Freeman et al., 2007). Brunello Cucinelli, for example, aligned with the precept that individuals should voluntarily work together "to create sustainable relationships in the pursuit of value creation" (p. 311) when he stated on his Web site under "My Creed":

> I dream about a form of humanistic modern capitalism with strong ancient roots, where profit is made without harm or offence to anyone, and part of it is set aside for any initiative that can really improve the condition of human life: services, schools, places of worship and cultural heritage.

Similarly, Eileen Fisher's corporate Web page devoted multiple subsections to their "Business as a Movement" philosophy. In their mission statement, the tenets of shareholder capitalism were presented in a bulleted list addressing "Individual Growth and Well-Being," "Joyful Atmosphere," and "Social Consciousness." Under this last heading they included the statement, "practice business responsibly with absolute regard for human rights." Stella McCartney phrased her brand's stakeholder capitalism initiatives as collective action using the pronoun "we":

> We understand that it is our responsibility to do what we can to become a more sustainable company. We are responsible for the resources that we use and the impact that we have. We take responsibility for operating a business and maintaining a supply chain that respects the planet as well as the people and animals on it.

Thus, through internal and external channels each brand positioned their companies as pursuing more than profit. It is notable that none of them mention their position as a luxury enterprise, focusing instead on the noble works and ideologies behind the product.

The head designer/creative director of a major luxury fashion brand is often cast as the lone creator of the branded goods regardless of their actual production role. Often, the numerous individuals working alongside the head designer are obscured to raise the one individual above the rest who receives acclaim and blame for the brand's direction through media discourse (Jones, 2015). In the discourse of luxury fashion, the ability to highlight an individual as "creator" offers a certain competitive advantage in the marketplace. External discourses outlined the internal structure of each of these companies as unique in a world that celebrates the lone creator.

A journalist describing the internal structure of Eileen Fisher used the analogy of "a family without parents" (Malcolm, 2013, September 23, p. 55). In business speak, this equates to a flat management style, long favored in tech industries as a mechanism to lead people from behind and allow them space to creatively problem-solve (Lemons, 2015, n.p.). Similarly, Cathy Horyn noted the "lack of hierarchy" at the Stella McCartney offices (Horyn, 2012, February 22, n.p.). The New Yorker was less kind in its description of Brunello Cucinelli's business: "He has enacted a peculiar fantasy of beneficent feudalism, with himself as the enlightened overlord, and the residents, many them his employees, as the appreciative underlings" (Mead, 2010, March 29, p. 72). Regardless of the analogy, the discourse of business organization and leadership style around each company was presented as unique and specific to each brand.

Internal and external discourse concerning the business practices of these three brands revealed two competing themes. In the text, there was oscillation between reverence for the creator (i.e., the founder) and a recognition of the collective efforts required to realize the ideological mission. The creator served as the source or keeper of the ideology, while those who labored for the creator were the realization of the ideology in practice. Mead (2010) noted a literal mimesis at the corporate culture of Brunello Cucinelli: "Although Cucinelli has no design training, his personal aesthetic has always informed the clothes, and those of his employees who wear Brunello Cucinelli to work end up looking a lot like Brunello Cucinelli" (p. 78). In an interview with Stella McCartney, it was her husband, Alasdhair Willis, who made the connection between the creator's ideology and the company's operations: "I think Stella is associated with a value system—sustainability, moral values—that's absolutely relevant to the times. The other brands are just beginning to catch on to her values, but they don't have the authenticity" (Horyn, 2012, n.p.). Fisher was very cognizant of her status while also downplaying its importance:

> I assume the reason you are interested in interviewing me goes beyond me. I sort of stand for a whole company, and I want to make sure that people are honored and that I don't say anything that offends anyone or that hurts anyone…I planted the first seed, and now I look around there's this amazing garden. But I'm just an ordinary person. (Malcolm, 2013, p. 58)

The ideological stance of each brand's creator has been central to building the connection between the brand and sustainability. Historically, the creator was also a significant contributor to the conception of the brand as luxury. Coco Chanel's glamorously portrayed life gallivanting among the wealthy continues to underline the brand's mythic stature today. However, as many luxury fashion houses have moved to a revolving door of creative directors (Jones, 2015), luxury is increasingly built through advertising and pricing structures.

In the case of these three brands, journalists employed intertextuality to position their subjects within the luxury fashion industry. Intertextuality is a phenomenon that constructs webs of meaning and enables communication by drawing on preexisting discourses tangential to the one at hand (Hodges, 2015). Intertextual elements helped draw the brands into a preestablished realm of luxury, whether luxurious living, luxurious consumption, or simply other brands with a better-known luxury pedigree. For example, Malcolm (2013) quoted Eileen Fisher as saying she "doesn't have a lot of needs" but pointed out in an adjoining parenthetical that Fisher "enjoys the privileges of the one per cent" and "It should be added that she is a political liberal" (Malcolm, 2013, September 23, p. 60). As a lesser known brand in the USA, Brunello Cucinelli was connected to the realm of other luxury brands when described as a producer of "clothing, which is favored by the kind of wearers who typically might choose Armani Black Label or Chanel for a formal occasion, such as meeting a President" (Mead, 2010, March 29, p. 72)

Brand-constructed discourse employed intertextuality in two ways: (a) drawing on preestablished tropes and general ethical values to bring their ideology into focus and (b) to bridge the ideology/object divide. The first instance of brand-constructed intertextuality was largely accomplished on the brands' "About Us" pages. Cucinelli's intertextual discursive practices come through the numerous quotes he shares from well-known philosophers—in fact, the title of his "About Us" page is "Philosophy." By referencing authors such as Kant, Dante, Deluze, and St. Francis of Assisi, the brand utilizes cultural capital as these authors signal a classical education—once only the purview of the wealthy. At the same time, "Philosophy" used the conceptual frameworks of these philosophers to create core values for Cucinelli's business ethos. This, then, established a specific social capital: "membership in a group" around shared ethical concerns. This granted Cucinelli's products a "credential" that exerted a "multiplier effect" as those in the group buy the goods based on "durable obligations subjectively felt" (Bourdieu, 1986, p. 22). In the end and regardless of values, brands are a business and this entire discursive exercise must achieve the necessary end of converting social capital into profit.

In external discourses, Stella McCartney is inevitably linked to her rock music lineage through her father, Paul McCartney. The persistence of this narrative for McCartney even after she had worked in the industry for fifteen years was illustrated when Horyn (2012) began her profile of McCartney by devoting several paragraphs describing McCartney's family attending her father's concert. However, McCartney's internal discourse never mentions this association. Instead, the intertextual elements included referenced personally acquired social capital and her sustainability credentials. McCartney's internal discourse connected her to the upper echelons of the luxury fashion hierarchy by highlighting her education at Central Saint Martins and her tenure at another luxury fashion house, Chloé. The brand also explicitly connected creator and ideology on the "About Stella" page stating, "Stella McCartney's commitment to sustainability is evident throughout all her collections and is part of the brand's ethos to being a responsible, honest, and modern company."

The ideological product

The systems employed for fashion production have become increasingly hidden from view, reducing oversight and aggregating externalities, particularly in terms of labor issues. As Sullivan (2016) noted through a Marxist critique, "At every step in its production, then, fashion exploits the labour of those involved in its complex and mostly opaque supply chain" (p. 40). Even luxury fashion goods moved from the ateliers to the factory as demand for couture dwindled and consumers flocked to the more affordable prices of RTW. Thus, one cannot assume that luxury goods are produced in superior conditions promoting worker welfare or environmentally responsible processing. In fact, a recent report from the nonprofit organization Fashion Revolution CIC (2017) indicated some of the most well-known luxury brands like Chanel, Hermès, and Dior offer little transparency to stakeholders. While these brands may detail commitments to operating a socially just and environmentally responsible company, if and how they satisfy those commitments remain hidden. Alternatively, some luxury conglomerates like Kering—parent company of Stella McCartney among others—have released their own sustainability reports detailing their initiatives on social and environmental fronts (Abnett, 2016, May 3). It becomes a natural point of difference when a brand's production process or product materiality strays from the basic and ventures into the novel. In this case, "novel" means environmentally and/or socially conscious sourcing for materials and labor. As established above, this is not the norm for most fashion brands, luxury, or mass market.

Discursively constructing the what and how of their products was an opportunity for the three brands to "ideologically square" (van Dijk, 2011, p. 397) goods against those that are neither luxurious nor sustainable. Ideological squaring produces an us versus them dichotomy; they (i.e., the rest of the fashion industry) may be producing their goods in an irresponsible manner but we (i.e., Fisher, Cucinelli, and McCartney, as brands) are doing everything we can to be ethical and thus substantially different. Jones and Hawley (2017) noted the reliance on ideological squaring in sustainable fashion discourse in Vogue magazine's editorial feature, "Style Ethics." The impression that luxury goods require high levels of artisanship, extended construction time, and extreme difficulty in production is woven into their product story and further heightened when an ethos of sustainability is introduced.

Fisher, Cucinelli, and McCartney used distinctly different discursive practices on their product pages. Fisher explicitly framed its apparel and accessories as sustainable by discussing environmentally sound production processes and highlighting the communities where they were produced. Cucinelli relied on cultural capital and price point to narrate their products, so artisanship and aesthetics were the subject matter, not the environment. McCartney utilized a combination of both approaches and—as the most aligned with the traditional world of fashion through education and branding—brought the novel and new into the conversation about luxury and sustainability.

Eileen Fisher's ideology was explicitly connected to product through descriptors such as "fair trade" and "organic." One sweater's production process was described as "Knit in an alternative supply chain designed around community-based workshops in Arequipa, Peru" detailing both process and place. Furthermore, consciously produced goods offered hyperlinks where a consumer could look "Behind the Label" and understand the meaning and processes of terms like "organic cotton." Here, videos, definitions, and cost analyses constructed a comprehensive corporate discourse on sustainable action, not just abstract beliefs or values. Noticeably absent, however, was any information addressing the presence of non-luxury, functional fibers in many of the yarn blends such as nylon and elastane, both of petrochemical sources. Still, Fisher operationalized its sustainability values down to the level of sustainable practices and attempted to show how its wares were a product of its ideology. Goods were imbued with an ethical credential, as Fisher's product page used, albeit limited, transparency to promote its values and differentiate its brand.

Comparatively, Cucinelli's product page did not narrate goods using the company's ideology as established in their "Philosophy" section or magazine editorials. While the descriptions for individual products highlighted traditional luxurious elements such as quality materials, inspiration sources, and place of production (e.g., Italy), they did not reveal sources of raw materials, human labor inputs, or environmental rigor of production. Instead, Cucinelli's discursive strategy was more poetic and raised issues of femininity, elegance, and fashion as a mode of being. In Cucinelli's world silhouettes were "renewed," there were "refined interpretations" of blouses, and a particular piece was described as "the dialogue between feminine allure and everyday functionality revisit[ing] the codes of luxury in a contemporary spirit." In this narrative, ethical values are implicit or presupposed while beauty, being, and becoming conjure an ontology of lived aesthetics.

McCartney's product page represented a blend of both Fisher's and Cucinelli's approach as they strategically decided when to invoke brand ideology and associated it with a specific product. Discursively, McCartney's product descriptions were about fashion first: design, inspiration, and the pedigree of style. The founder, traditionally educated in fashion, has been legitimated by the fashion media. Where production could potentially contradict the personal values of Stella McCartney, however, brand ideology was explicitly narrated into the product. Stella McCartney is a vegetarian so every shoe, for example, included the statement: "This item is made from non-leather, cruelty-free materials using highly skilled manufacturing techniques. This is part of our ongoing commitment to animal and eco-friendly fashion." McCartney's product page also highlighted organic cotton use. Unlike Fisher, the reviewed product descriptions did not comment on labor abstracting production in statements like "this knit cardigan in azure blue is crafted from felted wool for a soft yet dense feel." Abstracted language obscures social actors and process (Machin & Mayr, 2012). In this case, who did the "crafting" is abstracted, focusing instead on the product's superior nature. Discursively, McCartney's product page prioritized aesthetics and fashion while occasionally mentioning larger social values. In this way, McCartney's products exist in both realms of luxury fashion and ethical values with limited accountability or transparency regarding the latter.

Both Cucinelli and McCartney, then, separated their business ethos or brand ideology from the presentation and sale of their products. Through magazine editorials and "About Us" statements, these entities created ideological discourses where ethical and social concerns trump all things. Without consistently showing consumers how this ideology translated or operated within a product, however, the material consequences of these values remained ambiguous and open for interpretation. Following the logic of discourse as knowledge creation (Foucault, 1970), this also illustrated a lost opportunity for fashion brands to pass on sustainability-relevant knowledge to the consumer. This practice could also be defined as "greenwashing" (Lyon & Maxwell, 2011). If a consumer is aware of the general sustainable ethos of a company, the consumer could conceivably presume that all goods produced by the company follow that sustainable ethos. The same could be said of luxury goods. This lack of transparency could simply be a marketing strategy focused more on sales than production. At the same time, considering the ethical reputation these brands are attempting to establish, it also leaves too much room for misinformation, false assumptions, and outright deceit.

To narrate their products as luxuries, sustainable, socially responsible, or all three, each brand employed the concept of place. True to their discursive style, Cucinelli described "high quality Japanese produced new blue denim" to invoke the aesthetic and technological superiority of a denim pant. Fisher described jeans made in Los Angeles as part of the company's "efforts to support US manufacturing." McCartney described a sweater as "crafted from an Italian wool blend" to communicate quality but also reassured consumers that it was "responsible sourc[ed]." Geography and location were invoked as metonyms for qualities other than themselves and the actual people and ecosystems that subsisted there. There then exists the possibility of fetishizing, misunderstanding, or accepting place as an unqualified standard of quality, socioeconomic justice, or environmental practices. Because place has a central role in both sustainability and luxury, however, discussing the human geography of production and the sources of raw materials is a prerequisite for any fashion brand deemed sustainable luxury.

Hedging sustainability bets with low modality

While Fisher, Cucinelli, and McCartney largely took their status as luxury brands for granted, attempts to diminish the expectations of sustainability were found across discursive sites. Malcolm (2013) noted of Eileen Fisher's processes that "dyes that are not toxic are preferred if not always insisted on" and it "tries to be as green as it can without losing its shirt" (p. 56). Similarly, Cucinelli stated, "I would like to make a profit using ethics, dignity, and morals … I don't know if I'll be able to, but I'm trying" (Mead, 2010, March 29, p. 76). McCartney's corporate page ended the description of their CSR initiatives with "We will probably never be perfect, but you can rest assured that we are always trying." By employing words like "tries/trying" and "preferred," the discourse constructors revealed some of the ideological work at hand. In linguistic terms, it is described as "low modality" where the speaker or writer "might want to appear to be firmly aligned to an idea or thing but at the same time wish to limit how much this is represented in terms of a firm promise or command" (Machin & Mayr, 2012, p. 186).

The most explicit form of hedging came from McCartney's interview (Horyn, 2012, February 22). Horyn noted that McCartney did not advertise the ethics of her brand on her hangtags because (a) she wants the products to be appreciated for their design and (b) "the system of fair-trade, sustainable, ecological products is not perfect" (n.p). Of course, that interview was conducted at least five years ago and McCartney seems to embrace the ethics behind her label more explicitly now through her Web site. Sustainability is difficult to integrate successfully and is possibly unattainable within the current limits of capitalist enterprise. By employing low modality language, these brands cannot be faulted or criticized if they fail to meet the lofty ideals they have set. There appears to be no such fear of failure in meeting luxury expectations, however.

"Sustainable Luxury" Discursive Intersections

Despite their supposed dualities, this discursive analysis revealed spaces for luxury and sustainability to live in harmony. Based on this study, we propose the following illustration to outline the domains of these concepts (Fig. 2.1). The qualities central to both sustainability and luxury as discourses include materiality, quality, the maker, closely managed supply chains, and place (i.e., of production or source of materials). As the illustration implies, there are multiple elements for each domain that are not or cannot be found in the other. These excluded elements, particularly for sustainability, point to potential issues when combining the discourses. However, within the context of neoliberal, growth-oriented capitalism, sustainable luxury—sufficiently meeting desire—may be the limit of fashion's sustainability initiative.

Fig. 2.1 Intersecting discursive domains of luxury and sustainability

Conclusion

The three luxury fashion brands detailed above materialized their ideologies of selective sustainability across product categories, branding initiatives, and media discourse. The internally and externally constructed discourses served to position each brand among a web of signifiers connoting both luxurious and sustainable consumption, sometimes intersecting the two but with emphasis on the latter. All three brand discourses evoked stakeholder capitalism and engaged in ideological squaring to set themselves apart from other luxury brands. However, they also hedged through low modality to limit any expectations of their initiatives' impact. We noted a potential greenwashing effect when the brand ideology was taken for granted and not explicitly applied to products. Luxury was consistently taken for granted by all three brands with the most of the ideological work devoted to building an image of sustainable fashion.

The mere fact that Eileen Fisher, Stella McCartney, and Brunello Cucinelli are, each in their own way, forging a path for sustainable luxury companies is certainly noteworthy. However, it is also true that the path they are forging is one of "weaker sustainability" (Williams & Millington, 2004). From the above discussion, it can be surmised that both "luxury" and "sustainability" are actively evolving discourses, as separate domains and intersecting constructs. Fashion media and brand-constructed discourse are sites for discursive struggles where the definitions, limits, and forms of these concepts are constantly negotiated. With each editorial, product description, and corporate mission, magazine editors, online retailing merchandisers, and corporate public relation managers are shaping realities for their readers. They not only stimulate desire for their featured goods but also create a liminal space between the domains of "sustainability" and "luxury."

The potential implications here are twofold: (a) a space is carved out for the fashion industry to co-opt sustainability without imparting fundamental change to demand-driven consumption, and (b) sustainability becomes absorbed by the luxury discourse, inaccessible to all. Since sustainability will require collective action among the many, not the few, the second implication has the potential to stall sustainability as a global movement to enhance the lives of all citizens. Though some may argue for a "trickle down" approach here, slowly radiating out into less affluent communities, urgency is needed in several domains sustainable fashion is trying to address (e.g., waterway pollution, labor exploitation, and overconsumption). What remains is an uphill discursive struggle as the contradictions between the two domains remain unresolved.

References

Abnett, K. (2016). Kering goes public with sustainability report, revealing progress and pain points. *Business of Fashion*. Retrieved May 3, 2015 from https://goo.gl/gBextQ.

Bagwell, L. S., & Berhheim, B. D. (1996). Veblen effects in a theory of conspicuous consumption. *The American Economic Review*, *86*(3): 349–373. Retrieved from https://goo.gl/2cyBY7.

Berry, C. J. (1994). *The idea of luxury: A conceptual and historical investigation*. Cambridge: Cambridge University Press.

Bocken, N. M. P., & Short, S. W. (2016). Towards a sufficiency-driven business model: Experiences and opportunities. *Environmental Innovation and Societal Transitions, 18,* 41–61. https://doi.org/10.1016/j.eist.2015.07.010.

Bourdieu, P. (1986). The forms of capital. In J. Richardson (Ed.), *Handbook of theory and research for the sociology of education* (pp. 15–29). Westport, CT: Greenwood.

Chaffee, S. H. (1991). *Explication*. London: Sage Publications.

Cherny-Scanlon, X. (2016). Putting glam into green: A case for sustainable luxury fashion. In S. Dhiman & J. Marques (Eds.), *Spirituality and sustainability* (pp. 183–197). Switzerland: Springer International Publishing. https://doi.org/10.1007/978-3-319-34235-1_12.

Clark, H. (2008). Slow + Fashion—An Oxymoron—Or a promise for a future…? *Fashion Theory: The Journal of Dress, Body, & Culture, 12*(4): 427–446. https://doi.org/10.2752/175174108x346922.

Cochrane, L. (2015). Fashion world fears designer burnout as pressure takes its toll. The Guardian. Retrieved November 1, 2015 from https://goo.gl/jsPhTj.

Cooper, T. (2005). Slower consumption: Reflections on product life spans and the "throwaway society". *Journal of Industrial Ecology,* 9(1–2), 51–67. https://doi.org/10.1162/1088198054084671.

Crane, D., & Bovone, L. (2006). Approaches to material culture: The sociology of fashion and clothing. *Poetics, 34,* 319–333. https://doi.org/10.1016/j.poetic.2006.10.002.

Curwen, L., Park, J., & Sarkar, A. K. (2012). Challenges and solutions of sustainable apparel product development: A case study of Eileen Fisher. *Clothing and Textiles Research Journal, 31*(1), 32–47.

de Graaf, J., Wann, D., & Naylor, T. H. (2014). *Affluenza: how overconsumption is killing us—And how to fight back* (3rd ed.). San Francisco: Berrett-Koehler Publishers Inc.

Dobers, P., & Springett, D. (2010). Corporate social responsibility: Discourse, narratives and communication. *Corporate Social Responsibility and Environmental Management, 17,* 63–69. https://doi.org/10.1002/csr.231.

Dybicz, P. (2004). Ethical consumption within critical social policy. *Journal of Progressive Human Services, 15*(2), 25–43.

Fairclough, N. (2001). The discourse of new labour: Critical discourse analysis. In M. Wetherell, S. Taylor, & S. J. Yates (Eds.), *Discourse as data: A guide for analysis* (pp. 229–266). London: Sage Publications Ltd.

Fashion Revolution CIC. (2017). *Fashion transparency index: A review of 100 of the biggest global fashion brands and retailers ranked according to how much they disclose about their social and environmental policies, practices and impact.* Retrieved from https://goo.gl/iNbDGW.

Foucault, M. (1970). *The order of things: An archaeology of the human sciences.* New York: Random House.

Freeman, E. R., Martin, K., & Parmar, B. (2007). Stakeholder captialism. *Journal of Business Ethics, 74,* 303–314. https://doi.org/10.1007/s10551-007-9517-y.

Godey, B., Pederzoli, D., Aiello, G., et al. (2012). Brand and country-of-origin effect on consumers' decision to purchase luxury products. *Journal of Business Research, 65*(10), 1461–1470. https://doi.org/10.1016/j.jbusres.2011.10.012.

Heiderstadt, D. (2015) Fashion agenda: The goodbye game. *WWD, 210*(25): 8–10, 12, 14.

Hodges, A. (2015). Intertextuality in discourse. In D. Tannen, H. E. Hamilton, & D. Schiffrin (Eds.), *The handbook of discourse analysis* (pp. 42–60). Malden, MA: Wiley.

Horyn, C. (2012). What drives Stella McCartney. *The New York Times Magazine.* Retrieved February 22, 2012 from https://goo.gl/AnF6jn.

Jones, K. B. (2015). 'When one Dior closes': The discourse of designer changeovers at historic fashion houses. *Fashion, Style & Popular Culture, 2*(3), 321–331. https://doi.org/10.1386/fspc.2.3.321_1.

Jones, K. B., & Hawley, J. M. (2017). "Chic but scrupulous, down to the very last stitch": "Style Ethics" in American Vogue. *Fashion Practice, 9*(2), 280–302.

Jorgensen, M., & Phillips, L. J. (2002). *Discourse analysis as theory and method.* London: Sage Publications.

Kapferer, J.-N., & Michaut, A. (2015). Luxury and sustainability: A common future? The match depends on how consumers define luxury. *Luxury Research J, 1*(1), 3–17.

Kawamura, Y. (2005). *Fashion-Ology: An introduction to fashion studies.* Oxford: Berg.

König, A. (2006). Glossy words: An analysis of fashion writing in British Vogue. *Fashion Theory: The Journal of Dress, Body & Culture, 10*(1/2), 205–224.

Laclau, E. (2006). Ideology and post-Marxism. *Journal of Political Ideologies, 11*(2), 103–114. https://doi.org/10.1080/13569310600687882.

Lemons, J. F. (2015). *Flat management.* SAGE Business Researcher. Retrieved February 2, 2015 from https://goo.gl/MkFvLZ.

Lockwood, L. (2016). Special report: Instant fashion salvation or gimmick? *WWD, 211*(40), 8–11.

Luxury. (2010). In Stevenson, A., & Lindberg, C.(Eds.), *New Oxford American Dictionary.* Retrieved from http://www.oxfordreference.com/view/10.1093/acref/9780195392883.001.0001/m_en_us1264907.

Lyon, T. P., & Maxwell, J. W. (2011). Greenwash: Corporate environmental disclosure under threat of audit. *Journal of Economics and Management Strategy, 20*(1), 3–41. https://doi.org/10.1111/j.1530-9134.2010.00282.x.

Macdonell, D. (1986). *Theories of discourse: An introduction.* Oxford: B. Blackwell.

Machin, D., & Mayr, A. (2012). *How to do critical discourse analysis: A multimodal introduction.* London: Sage.

Malcolm, J. (2013). Nobody's looking at you: Eileen fisher and the art of understatement. *The New Yorker.* 52–63.

Mead, R. (2010). The prince of Solomeo: The cashmere utopia of Brunello Cucinelli. *The New Yorker*, 72–79. Retrieved March 29, 2010 from https://goo.gl/a1J1sE.

Princen, T. (2005). *The logic of sufficiency*. Cambridge, MA: The MIT Press.
Rose, G. (2001). *Visual methodologies: An introduction to the interpretation of visual materials*. Thousand Oaks, CA: Sage.
Silverstein, M. J., Fiske, N., & Butman, J. (2003). *Trading up: The new American luxury*. New York: Portfolio.
Solca, L. (2017). *Is Hermès drifting? Business of Fashion*. Retrieved February 14, 2017 from https://goo.gl/tLRXjf.
Sullivan, A. (2016). Karl marx: Fashion and capitalism. In A. Rocamora & A. Smelik (Eds.), *Thinking through fashion: A guide to key theorists* (pp. 28–45). London: I.B. Tauris.
Thomas, D. (2007). *Deluxe: How luxury lost its luster*. New York: Penguin Press.
van Dijk, T. A. (2011). Discourse and ideology. In T. A. van Dijk (Ed.), *Discourse studies: A multidisciplinary introduction* (pp. 379–407). London: Sage Publications.
Veblen, T. (2004). Conspicuous consumption. In D. Purdy (Ed.), *The rise of fashion: A reader*. Minneapolis: University of Minnesota Press.
Walker, S. (2010). Temporal objects—Design, change and sustainability. *Sustainability, 2,* 812–832.
Williams, C. C., & Millington, A. C. (2004). The diverse and contested meanings of sustainable development. *The Geographical Journal, 170*(2), 99–104. https://doi.org/10.1111/j.0016-7398.2004.00111.x.
World Commission on Environmental Development (WCED). (1987). *Our common future*. Oxford: Oxford University Press.

Chapter 3
The Marketing of Sustainability and CSR Initiatives by Luxury Brands: Cultural Indicators, Call to Action, and Framework

RayeCarol Cavender

Abstract This conceptual study addresses the sustainable development (SD) and corporate social responsibility (CSR) initiatives of global luxury brands and their lack of information dissemination on these endeavors to the luxury consumer market. Strategies for global luxury companies to connect their consumers to the sustainability messages that underlie their core values and inform their business operations are still nascent in the luxury brand management literature. A case study tracing the SD and CSR initiatives of leading luxury conglomerate, LVMH, since the inception of its environmental department (i.e., 1992–2017) was conducted alongside an in-depth historical review of the larger luxury goods industry. This approach allowed for concurrent examination of LVMH's sustainability initiatives over time and within the context of the evolving luxury goods environment (e.g., changing luxury consumer motivations and values). Data were collected from primary and secondary data sources. Two theoretical foundations, experiential marketing (e.g., Atwal & Williams, 2009; Pine & Gilmore, 1998) and sustainable luxury value (Hennigs et al., 2013), shaped the study. Findings revealed a conceptual framework for sustainable luxury brand communication (SLBC) that can serve as a guide for global luxury brands with sustainability programs in varying stages of development. The SLBC framework proposes the order in which brands communicate their SD and CSR initiatives to the luxury market, with the goal of educating and helping fuel the shift in consumer values toward a more sustainable paradigm. This research contributes to the growing body of literature on luxury brands' sustainability programs and their impact at the consumer and corporate-level.

Keywords Luxury · Sustainable development · Corporate social responsibility Strategic management

---

R. Cavender (✉)
University of Kentucky, 318 Erikson Hall, Lexington 40506, USA
e-mail: rayecarol.cavender@uky.edu

© Springer Nature Singapore Pte Ltd. 2018

C. K. Y. Lo and J. Ha-Brookshire (eds.), *Sustainability in Luxury Fashion Business*, Springer Series in Fashion Business, https://doi.org/10.1007/978-981-10-8878-0_3

Introduction

The clothing industry ranks just behind oil as the world's second most polluting industry (Muratovski, 2015). Exploitation of resources and people on the supply side is met with rampant overconsumption, fueled by the fast-fashion industry, on the consumer side (Ertekin & Atik, 2015). Even more startling, the World Wildlife Foundation estimates that without far reaching measures for change at the institutional and consumer level, "by 2030 we would need the equivalent of two planets' productive capacity to meet our demands" (Lovegrove, 2011, p. 7). However, research suggests that the emergence of sustainable fashion movements (e.g., slow fashion, conscious consumption) indicates that society as a whole is beginning to question the consumption-oriented dominant social paradigm, signaling an impending shift toward more sustainable consumption and increased commitment to sustainability initiatives by corporations (Bendell & Thomas, 2013; Ertekin & Atik, 2015; Grail Research, 2010; Hennigs, Wiedmann, Klarmann, & Behrens, 2013).

Sustainable development (SD) is "development that meets the needs of the present without compromising the ability of future generations to meet their own needs" (Kapferer & Michaut, 2015, p. 5). It is encompassed by the broader framework of corporate social responsibility (CSR) and comprises three pillars: ecological, slow growth, and social harmony. SD has been "identified by luxury specialists as one of the main strategic challenges and opportunities for the industry" (Bendell & Thomas, 2013, pp. 15–16).

Further, SD is now increasingly referenced as being at the top of luxury brands' corporate agendas, demonstrating a quick evolution from being considered an optional to an obligatory business activity (Ivan, Mukta, Sudeep, & Burak, 2016). However, some luxury groups, such as Louis Vuitton Möet-Hennessy (LVMH) who was the first company to establish an environmental department in 1992, have long since prioritized SD as a major operational activity across their organizations' entire value chain (i.e., sourcing, creating, manufacturing, logistics, distribution, marketing, servicing, waste, recycling). These firms consider SD from the perspective of the triple bottom line (i.e., social, environmental, economic) and as both a moral and economic imperative (Kaperer 2010; Rahman & Yadlapalli, 2015).

Luxury brands, traditionally adverse to marketing, have in recent years, increasingly engaged with consumers across multiple channels and touchpoints in response to the digital economy and experiential marketing orientation. However, many luxury brands have yet to incorporate narratives and stories that highlight sustainability initiatives into their brand communication strategies (Kapferer & Michaut, 2015). Ivan et al. (2016) stated that the "practice of sustainability require[s] early adopters to use the communication channels of the luxury business over time to be able to bring change to the social system," (p. 30) while Rahman and Yadlapalli (2015) claimed that "[r]esponding to changing social and cultural trends creates legitimacy for a luxury brand" (p. 193). Hopkinson and Cronin (2015) suggest "that long-term change, such as that needed for sustainability, arises through the linked actions of organizations and consumers" (p. 1384). Therefore, luxury groups' lack of communication on SD and CSR initiatives is arguably a missed opportunity to cultivate awareness and understanding at the consumer level and illustrate best practices for SD and CSR that can be emulated by companies in the wider apparel industry, two critical steps in repositioning the fashion industry toward a more sustainable paradigm.

Conceptualizing Sustainable Luxury Brand Communication

This study defines sustainable luxury brand communication (SLBC) as the strategic dissemination of SD information by luxury brands across channels (i.e., Web site, in store, mobile, social) to experientially engage with and inform luxury consumers about a company's internal SD and CSR initiatives, external collaborations and partnerships, and the environmental impact of the apparel industry. The term "brand communication" is adopted as opposed to "marketing" as research suggests that successful sustainability marketing messages make subjective appeals that downplay commercial motives (Dach & Allmendinger, 2014; Kapferer & Valette-Florence, 2016). This type of communication is in line with current luxury marketing strategies that are not directly aimed at selling products, but instead promote luxury brands' essence and dream value (Kapferer, 2015).

Strategies for global luxury players to connect their consumers to sustainability messages that underlie their core values and coincide with initiatives being executed through their business operations are limited. Further, the potential benefits of this information sharing with target consumers and the general public are still nascent in luxury brand management research. The purpose of this conceptual study is to address this gap by proposing a framework for brand communication by luxury companies on their SD and CSR initiatives.

Review of Literature

Sustainable Development and Luxury: Fundamentally Linked

A range of research (e.g., Fionda & Moore, 2009; Ivan et al., 2016; Kaperer, 2010; Ranfagni & Guercini, 2016; Vigneron & Johnson, 2004) has identified components of the luxury brand (e.g., product production integrity, quality, scarcity, beauty, durability) that suggest a symbiotic relationship with SD. Because luxury brands sell rare, resource-dependent products, a commitment to SD is intrinsically linked to brands' long-term operational success (Joy, 2015; Kaperer, 2010). Gardetti (2016) stated that "[s]ustainable luxury promotes the return to the essence of luxury with its ancestral meaning, the thoughtful purchase, the artisan manufacturing, the beauty of materials in its broadest sense, and the respect for social and environmental issues" (Gardetti, 2016, p. 2). Sustainable luxury not only cultivates heightened respect for the environment and social development, but is also synonymous with social development, cultural heritage, and art patronage (Gardetti & Giron, 2014).

Heritage luxury brands have had to become adept at crafting a careful balance between tradition and innovation in order to maintain relevancy in the market over time. This includes, on one hand, sustaining the legacy of local craftsmanship (i.e., specialized skills, training) and savoir-faire that is representative of brands' countries of origin (Joy, 2015). Luxury leaders in SD must also continue to invest in technology and R&D to develop innovations in materials and production processes geared toward sustainable solutions and protecting their core businesses (Carcano, 2013). A commitment to SD also calls for clear traceability and transparency within the supply chain (Gardetti & Giron, 2014). This is an advantage for true luxury companies (e.g., Louis Vuitton) as they are already highly vertically integrated and have tight holds on their sourcing, production, logistics, and distribution (Kaperer, 2010; Rahman & Yadlapalli, 2015). In recent years, these companies have further consolidated their supply chains by acquiring suppliers of the scarce raw materials used in their product lines. LVMH initiated this industry trend by acquiring Heng Long reptile skin tannery in 2011 and leather tannery, Tanneries Roux, in 2012 (Safe, 2017).

Sustainable Development as a Core Luxury Value

Research demonstrating an alignment of sustainability drivers with the luxury brand ethos is supported by the growing commitment to sustainable development within the luxury goods industry. Successful brands in this sector leverage their know-how in the areas of innovation, creativity, and excellence to provide "the best products and experiences for well-informed, educated, sophisticated, demanding and discerning clients" (Gardetti & Giron, 2014, p. 20). Further, strategic management of a luxury brand requires a long-term vision with measurable strategies that consider two main dimensions, the economic and the symbolic (i.e., sociocultural context of reference; Carcano, 2013). As early as 1992, future-oriented luxury groups like LVMH began infusing SD into their brands' DNA in anticipation of industry growth, as a social imperative, and as a business imperative to preserve the rare natural resources (e.g., skins, leathers, pearls) that must be protected to sustain long-term growth (Ranfagni & Guercini, 2016). For these companies, explicating SD as a core value has meant establishing environmental departments and steering committees that prioritize SD objectives, voluntarily submitting to internal and external auditing on SD measures, and including auditing results in the companies' annual reports. Bendell and Kleanthous' (2007) Deeper Luxury Report for the World Wildlife Fund was a call to action for luxury groups to report annually on the

Global Reporting Initiative (GRI) guidelines and marked a shift toward increased transparency and commitment to SD in the sector. Because of global initiatives such as the GRI guidelines and ISO 26000 standards, sustainable reporting is now "a main stream business activity for companies" (Rahman & Yadlapalli, 2015, p. 195).

Many "born sustainable" companies (e.g., Elvis and Kresse, The Summer House), positioned as "luxury brands," have also emerged in recent years and are achieving success by placing sustainability issues at the heart of their raison d'etre and operational strategies (e.g., creating products from waste materials). These companies are challenging "assumptions about the nature of luxury" and helping to elevate the sustainability discourse (Bendell & Thomas, 2013, p. 22). However, these smaller "elegant disruptors" are niche business that do not have the commercial reach or brand recognition of the larger luxury houses, nor the capital to initiate large-scale change throughout entire supply chains, within communities, or through investment in R&D.

As purveyors of the cultural zeitgeist with high brand visibility, luxury brands with SD as a core value (e.g., Fendi, Loro Piana) are poised to share about their SD initiatives more regularly through innovative communication efforts that educate and inform their consumers, who themselves often serve as change agents and opinion leaders in consumer society. Although brands with SD as a core value may be better positioned to qualify the meaning and importance of sustainability, luxury brands that have only begun investing in SD more recently can still engage in SLBC (Ivan et al., 2016). This means conveying that "a step by step progressive modification and control is taking place with high goals at the end" (Kapferer & Michaut, 2015, p. 15). In this context, more luxury brands can begin to facilitate the sustainability discourse by highlighting the intrinsic attributes (e.g., rarity, durability) that make investing in luxury products a reasonable alternative to mass market consumption (Janssen, Vanhamme, Lindgreen, & Lefebvre, 2014).

Theoretical Foundation for Guiding the Study

Two theoretical foundations, sustainable luxury value and experiential marketing, guided the development of the SLBC framework.

Sustainable Luxury Value

There is a range of research (e.g., Atwal & Williams, 2009; Berthon, Pitt, Parent, & Berthon, 2009; Beverland, 2006; Kapferer & Bastien, 2009; Phau & Prendergast, 2000; Vigneron & Johnson, 2004) that conceptualizes luxury value from both the company and consumer perspective, and more recently, a growing body of literature that confirms the fundamental link between many intrinsic luxury values and sustainability (e.g., Gardetti, 2016; Joy, 2015). Although the connection between the luxury essence and SD is now widely accepted at the organizational level, there are still mixed research findings about consumers' perception of this relationship. This may explain why research that operationalizes sustainable luxury value is still nascent in the literature. However, the Hennigs et al. (2013) sustainability diamond does propose the dimensions of sustainable luxury value and, thus, serves as a guiding framework to the present study. Drawing from literature on luxury brand management and luxury consumer value, (Hennigs et al., 2013) proposed a comprehensive, multidimensional framework that explicates key dimensions of value-based social and environmental luxury excellence. For the proposed dimensions on which consumers evaluate the ethical performance of luxury brands (i.e., financial, functional, individual, social), the authors state that in order to maximize "customer perceived value in the context of sustainability excellence, each management decision has to be reflected from the customer's perspective and the meaning that consumers attach to the multifaceted product attributes" (Hennigs et al., 2013, p. 33).

The financial value dimension involves the monetary value of luxury, including what is given up to obtain the product (Hennigs et al., 2013). Attributes of sustainable luxury (e.g., preservation of scarce materials, timelessness) can be leveraged to further substantiate the high product costs that offset a company's socially and environmentally responsible manufacturing. The functional value dimension links the company's sustainable production processes (e.g., ethically sourced raw materials, skilled artisans, bespoke craftsmanship) to material attributes (e.g., quality, durability, uniqueness) of sustainable luxury that create value for the consumer (Hennigs et al., 2013). The individual value dimension considers consumers' personal orientation toward luxury. This includes the symbolic value of luxury products that consumers transfer to their own identities (e.g., ethical consumption) and the experiential value (e.g., sensations, feelings, cognitions, behavioral responses) derived from the total brand experience (Berthon et al., 2009). The social value dimension refers to the social collective and the symbolic value that a luxury brand signals to others (e.g., prestige). The current reorientation of luxury meaning is already expanding these signals to include social and environmental consciousness (Joy, 2015; Kapferer & Michaut, 2015). Social value also encompasses the experiential value of connection with a larger group (e.g., environmentally aware luxury consumers), trend, or cultural movement (Schmitt, 1999). It is notable that experiential value has a developmental aspect that matures with time and experience (Berthon et al., 2009). Thus, it is feasible that luxury companies' SD- and CSR-related brand communication will strengthen the individual and social value dimensions as consumer awareness of the environmental impact of apparel production and consumption continues to increase (Hennigs et al., 2013).

Experiential Marketing Orientation

In the digital age, consumers desire continuous experiential content across all channels and touchpoints of brands with which they engage. Because luxury goods are experiential in nature, luxury companies are in a unique position to apply experiential principles to their marketing strategies (Atwal & Williams, 2009). Experientially engaging consumers through SD- and CSR-related communication can facilitate the deep bonds between product and user that characterize "deeper luxury" and represents an additional topic area companies can draw from to create branded content (Ivan et al., 2016; Janssen et al., 2014).

Schmitt (1999) identified consumers as "rational and emotional human beings who are concerned with achieving pleasurable experiences" and categorized experiential marketing according to five different types of experiences, each of which has relevancy to SLBC (p. 53). Sensory experiences add value to products and brands through appealing to the senses. Luxury brands already facilitate sensory experiences through aesthetically beautiful, avant-garde content that nurtures their dream value (Kapferer, 2015). Sensory experiences can be leveraged through SLBC to expand the luxury dream to account for environmental concerns (Joy, 2015).

Affective experiences cultivate in consumers a desired emotional state (e.g., pride, empathy) but may pose a challenge when disseminated to a cross-cultural audience (Schmitt, 1999). Global luxury consumers come from markets of varying maturity and represent diverse cultural values, yet recent international research supports a growing sensitivity to sustainability among luxury shoppers worldwide. This suggests that SLBC is relevant to the global, cross-cultural luxury market (Kapferer, 2015).

Creative cognitive experiences challenge the intellect, often through surprise or provocation, and can be employed to address consumers' convergent or divergent thinking (Schmitt, 1999). Providing consumers with specific information on the environmental impact of the apparel industry, from raw material to product disposal (i.e., cradle to grave), and on the innovative solutions being implemented by luxury brands to foster more sustainable operations may equip consumers with the knowledge to better understand the impact of their own consumption habits.

Schmitt's (1999) fourth type of experiential marketing (i.e., physical, lifestyle, behavioral experiences) is activated to show consumers "alternative ways of doing things … alternative lifestyles and interactions" (p. 62). Through SLBC, companies that have recently accelerated their sustainability initiatives (e.g., Prada) can share the impetus for their strategic shift in transparent terms, acknowledging that SD is both an economic opportunity and necessary to preserve resources for future generations. This informative content may elicit the motivated and inspired responses that Schmitt (1999) suggests can prompt consumers' lifestyle and behavioral changes. Further, luxury consumers may become more aware of the disconnect that researchers have identified between positive perceptions of companies' sustainability initiatives and actually adopting sustainable behaviors in their own consumption practices and daily lives (Joy, 2015; Muratovski, 2015).

Schmitt's (1999) final experience, the social identity experience, highlights the individual need to feel a connection with reference groups and broader societal phenomena.

Pine and Gilmore (1998) explicated the four realms of an experience to be escapist, entertainment, educational, and aesthetic and varying across two dimensions: the degree of participation required (i.e., active or passive) and the connection (i.e., absorption or immersion) that unites the customer with the experience. Atwal and Williams (2009) adapted these dimensions to be involvement, or the degree of inter-activity between customer and supplier, and intensity which is the "perception of the strength of feelings toward the interaction" (p. 342). Similarly, consumers can determine their desired level of engagement with brands' SLBC out of the content available to them across channels and touchpoints, co-creating with the brand the "special and authentic experiences of connection" that luxury consumers have come to expect (Hennigs et al., 2013, p. 31).

Combined Theories for Sustainable Luxury Brand Communication Framework

The Hennigs et al. (2013) sustainability diamond links the drivers of sustainable luxury value at the organizational and consumer levels. While research demonstrates the importance of the experiential component to luxury value creation, it is evident that this component also underlies sustainable luxury value creation (Berthon et al., 2009; Kapferer, 2015). Experiential marketing orientation is also highly aligned with best practices for luxury marketing that downplay commercial motives while facilitating meaningful brand experiences for consumers across available channels (Dach & Allmendinger, 2014; Kapferer & Valette-Florence, 2016; Schmitt, 1999). When examined together, sustainable luxury value and experiential marketing orientation provide information that can inform the development of companies' SLBC strategies that both reinforce sustainable luxury value and illuminate the commitment to SD and CSR through experiential (e.g., educational) messages (Ivan et al., 2016).

Methods

An inductive research approach using grounded theory methodology was adopted to achieve the study's purpose of proposing a conceptual framework for brand communication by luxury companies on their SD and CSR initiatives. "A grounded theory approach focuses on the systematic gathering and analysis of data to derive theory and has been widely used in marketing and supply chain research" to explore phenomena in relatively early stages of development and/or to explore complex issues (Closs, Speier, & Meacham, 2011, p. 103). The grounded theory approach requires a rigorous coding method to help researchers develop their interpretations and is useful for building theory based on data collected from the field of study (e.g., document analysis; Closs et al., 2011).

Sample

To inform the study of luxury companies' SD and CSR, a case study was conducted tracing the sustainability initiatives of leading luxury conglomerate, LVMH, since its environmental department was established (i.e., 1992–2017). LVMH, with its portfolio of 59 brands, has the most mature (est. 1992) and comprehensive sustainability program in the luxury industry, with brands' strategic business decisions being driven through the ambitious LIFE (LVMH Initiatives for the Environment) program (Winston, 2017). LVMH's sustainable supply chain is often referenced as an exemplar (e.g., traceability, raw material conservation, supplier partnerships) in sustainable luxury discourse, but LVMH's ongoing commitment to managing its carbon footprint has garnered attention from wider industry sectors, governments, and NGO's (Carcano, 2013; Gardetti, 2016; Grail Research, 2010; Kaperer, 2010; Rahman & Yadlapalli, 2015). For example, in 2015, LVMH implemented an internal carbon fund, joining the select few companies (e.g., Disney, Microsoft) outside of the fossil fuels or auto industries that have self-imposed a carbon tax on their operations (Winston, 2017).

A more extensive view of SD and CSR in the luxury industry and the effect of a multiple case study was achieved through reviewing the SD and CSR initiatives of the 59 brands in the LVMH portfolio that represent five operating groups (i.e., fashion and leather goods, watches and jewelry, perfume and cosmetics, wine and spirits, selective retailing).

Data Collection

This study employed documentary research with literature review and case study analysis. This combination of techniques resulted in a rich database that informed the research at the company and consumer levels of the industry as both perspectives were considered central to the development of the SLBC framework. A comprehensive review of academic literature, encompassing both a company and consumer focus, was conducted in the areas of luxury brand management, sustainable luxury, and luxury consumption to gain a broad understanding of sustainability drivers and challenges in the luxury goods sector and the larger apparel industry.

This approach utilized primary and secondary data sources and was conducted with an in-depth historical review of SD and CSR activity in the luxury goods industry. This allowed for concurrent examination of LVMH and its brands' initiatives and communication on those initiatives over time within the context of the evolving luxury goods environment (e.g., changing luxury consumer motivations, values). Relevant articles from trade publications (e.g., *Luxury Daily, Business of Fashion*) were collected that provided current and archival industry information as well as company-specific information. News publications (e.g., *Wall Street Journal*) and business magazines (e.g., *Forbes*) were surveyed for information related to sustainability in the luxury sector, the broader apparel industry, and the sample company, with additional information being obtained from business databases (e.g., Mintel). Company documents (e.g., LVMH Sustainability Report from 2002 to 2016) and in-depth financial reports (i.e., annual results) were obtained from the parent website. Web sites and available social media channels of brands in the LVMH portfolio were reviewed to gather information on SD and CSR activities and communication efforts at the company level.

Data Analysis and Framework Development

According to best practices for inductive research, data analysis was approached in an emergent way, minimizing the potential for categorization bias as a priori themes were not developed (Closs et al., 2011). The researcher first used inductive content analysis to code the collected data and derive emergent concepts. Three themes emerged that are perceived to be cultural indicators of the state of sustainable luxury (i.e., macro- and micro-level factors). The cultural indicators are explicated and supported with examples from the data and employed in the development of the framework as they offer meaningful insights for companies' SLBC strategies. The indicators were combined with theoretically derived concepts (e.g., sustainable luxury value, experiential marketing) to guide framework development.

The resulting framework links luxury companies' sustainability initiatives to the multi-dimensional construct of sustainable luxury value and offers a hierarchy that prioritizes the order in which luxury brands should implement brand communication on their SD and CSR initiatives. To further explicate the framework, best practices in SD and CSR by LVMH, derived from case analysis, are discussed to offer examples for how SLBC can be disseminated to inform, educate, and increase awareness of SD and CSR among luxury consumers. To improve validity of the study, the researcher triangulated the study through the use of multiple data sources and employed constant comparative analysis and recontextualization techniques in order to capture the full complexity of SLBC. Two graduate students that were familiar with the research participated in peer debriefing to confirm the emerging concepts.

Results

This section first reviews the cultural indicators that emerged during content analysis and concludes with a discussion of the SLBC framework (i.e., antecedents, hierarchy).

Cultural Indicators of a Tipping Point for Sustainable Luxury Brand Communication

Cavender and Kincade (2015) conceptualized the luxury brand's strategic management response to environmental determinants (e.g., sociocultural environment) and the zeitgeist as a central function of successful luxury brand management. Three such environmental determinants, or cultural indicators, emerged from inductive content analysis that necessitates a strategic management response from luxury brands. These indicators of the sustainable luxury environment support the need for luxury leaders in SD and CSR to develop strategies for brand communication on these endeavors across channels and customer touchpoints in order to maximize sustainable luxury value (Cavender & Kincade, 2015; Hennigs et al., 2013).

Alignment with Emerging Consumption Paradigms

"Change is beginning to happen at both institutional and consumer levels. Individuals and society as a whole have begun to question the prevailing consumer culture" (Ertekin & Atik, 2015, p. 63). In response to overconsumption and growing environmental concerns, movements such as slow consumption, slow fashion, and conscientious consumption have emerged with the premise of consuming lower quantities of apparel by purchasing higher quality items (Joy, Sherry, Venkatesh, Wang, & Chan, 2012; Joy, 2015). On the supply side, these paradigms are referred to as slow design and are characterized by sustainable operations. Slow design fosters deeper bonds between brand, product, and user and "is a beautiful echo of the brand DNA of luxury products and hence cements the idea of sustainable slow fashion in luxury" (Ivan et al., 2016, p. 32). This alignment conveys the idea that, in contrast to many consumers' understanding of luxury as frivolous, the enduring quality of luxury products actually "encourage[s] more reasonable, responsible consumption and help[s] protect natural resources" (Janssen et al., 2014, p. 3).

Recent examples of paradigm shifts in consumer sectors such as food and cosmetics have resulted in increased transparency by companies regarding production methods, ingredients, etc. (Ertekin & Atik, 2015; Gardetti & Giron, 2014; Joy, 2015). In these cases, companies' employed marketing efforts to inform consumers of changes that were made to their product offerings (e.g., non-GMO, no animal testing on cosmetics). Just as consumers want assurance that certain ingredients are not in the foods they eat or cosmetics they use, they may also be interested to learn about luxury firms' R&D efforts that have resulted in more sustainable processes and materials. For example, LVMH-owned Loewe recently developed an adhesive that is now used in place of glue (an unsustainable material) in the production of its handbags ("Environmental Report," 2017).

Alignment with Shifting Consumer Values

Bain predicts that, by 2025, Millennials and Generation Z will account for 45% of the personal luxury goods market (Solomon, 2017). These highly digital generations value experiences over things, desire consumption experiences that allow them to express their deepest values, and seek personalized, emotional and frequent engagement with brands across channels. Like Baby Boomers before them, Millennials are "influencing the evolution of the whole luxury customer base" (Solomon, 2017, p. 9) and signaling an end to consumer marketing as it has long been known. This is necessitating a fundamental rethink of marketing's purpose (Muratovski, 2015). Companies must now convince "Millennials that they are 'doing good' when they purchase its brands [and]...demonstrate through their values, heritage, and meaningful actions that they help those in need, are socially responsible, are good environmental stewards, protect personal data, or are transparent and sincere" (Boston Consulting Group, 2013, p. 6).

Shifts in consumer values are also occurring in emerging markets as consumers are transitioning out of the discovery phase, characterized by conspicuousness and compulsive accumulation of luxury, and beginning to seek out experiential value in luxury consumption. Tastes and preferences of luxury consumers in non-Western markets "have been reportedly 'maturing' quicker than it took in the West, with more consumers focusing on aspects of the product and company beyond the visible appearance of the brand" (Bendell & Thomas, 2013, p. 15). Recent international research found that affluent buyers' sensitivity to SD was higher in China and Brazil than in the USA, France, Germany, or Japan. This suggests that "sensitivity to green caveats is now growing among the affluent buyers of the world" (Kapferer, 2015, p. 36) and is consistent with literature supporting the need for SLBC to educate and inform consumers in both mature and maturing luxury markets (Cervellon, 2013; Gardetti & Giron, 2014; Kapferer, 2015; Kapferer & Michaut, 2015; Kapferer & Valette-Florence, 2016).

Limited Consumer Understanding of Sustainable Luxury

The still limited mention of sustainability as a selection criterion for luxury product purchases and brand loyalty may be because a commitment to SD and CSR by luxury brands is a latent expectation for luxury consumers (Kapferer, 2015). That is, consumers believe that sustainability is intrinsically linked to the luxury business model (Achabou & Dekhili, 2013; Davies, Lee, & Ahonkhai, 2012; Kapferer & Michaut, 2015). Research also suggests that consumers' subjective definition of luxury affects whether they view luxury and sustainability as being compatible (Kapferer, 2015). Janssen et al. (2014) found that consumers' perceptions of fit between luxury and sustainability was higher for enduring products produced from scarce materials (e.g., jewelry) than for ephemeral products such as clothing items. Similarly, Cervellon (2013) found that consumers' considered craftsmanship, use of rare materials, and country-of-origin to signify sustainable luxury, yet only viewed brands' support of philanthropic causes as sustainable when there was a clear link between the brand and cause and when the associated marketing message was devoid of commercial purpose. These findings are particularly noteworthy because the still limited SLBC, even by luxury leaders in SD, tends to focus on CSR initiatives and partnerships (e.g., Louis Vuitton + Unicef campaign). One might argue that increasing brand communication on other SD commitments (e.g., supply chain, environment) may lend credibility to brands' communication about their philanthropic endeavors as consumers gain a more comprehensive understanding of companies' SD commitments.

In contrast to Cervellon (2013), Kapferer and Michaut (2015) found that consumers who define luxury as expensive or rare perceive a higher contradiction with sustainability than those who view luxury as extreme quality, and furthermore, that younger consumers (i.e., 18–34) are more likely to perceive a significant contradiction between sustainability and luxury than older consumers (i.e., 55–75). This demonstrates the need for increased awareness about brands' SD initiatives among younger generational cohorts (i.e., Millennials, Generation Z) because there is evidence that younger generations value sustainable business more than older generations (Boston Consulting Group, 2013; Solomon, 2017). Successful, targeted marketing and consumer education about brands' sustainable development principles and the criteria that make products sustainable (e.g., quality, durability) may be useful to address the misalignment of Millennials' values and their perceptions of luxury brands on SD factors.

Kapferer and Michaut (2014) observed a contradiction between consumers' perceptions of luxury consumption and sustainability when they perceived luxury to be superficial, yet findings also indicated that consumers' perceived luxury as promoting "true values." This suggests that luxury brands "need to promote their true values credibly to consumers. If consumers cannot perceive how the true values of luxury are in line with sustainable development principles, they continue to perceive only superficiality in the industry" (Kapferer & Michaut, 2014, p. 15).

Another possible explanation for mixed findings related to luxury consumers' perceptions of sustainability is that consumers only have a nominal awareness of the breadth of issues that SD entails and have very little understanding about the scope of sustainability within a brand's supply chain (Grail Research, 2010; Ivan et al., 2016; LaRocca, 2014). For example, Faust (2013) explored luxury consumers' perceptions of cashmere, finding a significant lack of consumer knowledge about the fiber. After education on the cashmere production process (i.e., from goat to garment), study participants cited increased understanding of the provenance of cashmere and suggested ideas for how luxury brands can communicate with consumers about the bespoke materials and sustainable production processes used to create their product assortments (e.g., videos playing in store). Faust concluded that "it seems obvious that articulating the story behind luxury goods would enhance the shopping experience and provide a sustainable competitive advantage for retailers (2013, p. 973).

An opportunity exists for luxury companies to leverage their high brand visibility, through strategic communication efforts, to frame and elevate the sustainability topic among target consumers and in wider consumer markets. Educating consumers on how they are furthering the SD agenda within their own operations, and the needed initiatives to counteract the environmental impact of the global apparel industry, may strengthen luxury brands' image as many consumers now consider SD to be a latent expectation of luxury yet are largely unaware of the scope of companies' SD activities (Kapferer & Michaut, 2015).

Sustainable Luxury Brand Communication Framework

The SLBC framework is situated within the context of the luxury goods environment and includes the cultural indicators that emerged from content analysis. The framework links luxury companies' sustainability initiatives to the multi-dimensional construct of sustainable luxury value (Hennigs et al., 2013) and offers a hierarchy (see Fig. 3.1) that prioritizes the order in which luxury brands can implement brand communication on their SD and CSR initiatives. Best practices in SD and CSR by LVMH are discussed to offer examples for how information on SD strategies can be disseminated through brand communication to inform, educate, and increase awareness among luxury consumers.

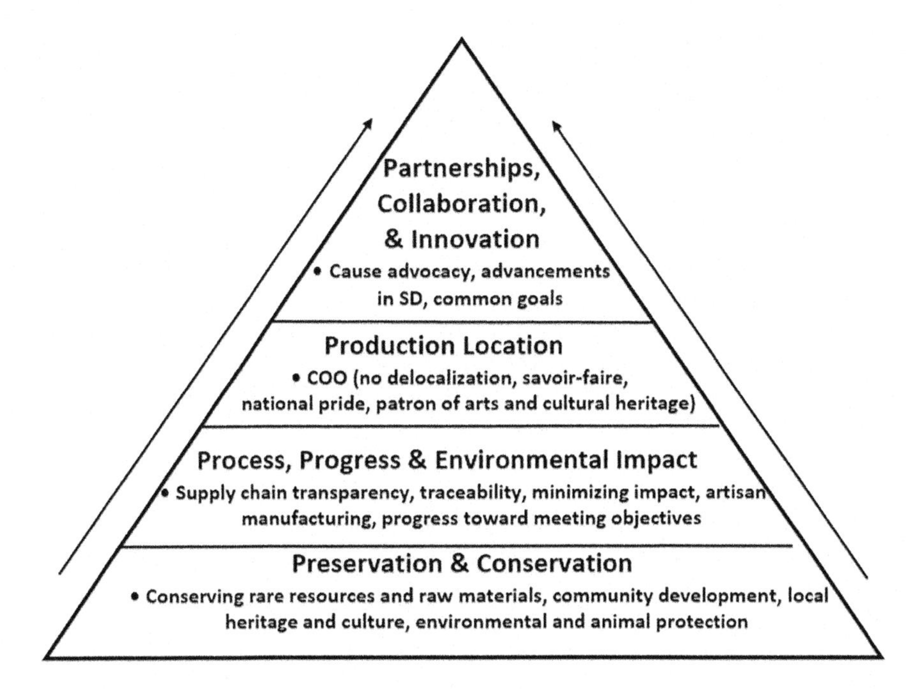

Fig. 3.1 Sustainable luxury brand communication (SLBC) hierarchy

Antecedents to Sustainable Luxury Brand Communication

Three antecedents emerged from content analysis that are integral to the development of SLBC strategies. Whether a company intends to communicate on a range of its environmental and social initiatives or prioritizes communication on a select few sustainability commitments (e.g., supply chain), it must infuse its authenticity, legitimacy, and transparency into the communication strategy (Freire, 2014; Gardetti, 2016; Guercini & Ranfagni, 2013; Ivan et al., 2016).

Although transparency is presumably inherent to a luxury company's SD and CSR brand communication strategy, research suggests that sharing both positive and less favorable information can lend credibility to a company's message. For example, companies can illuminate the areas in which they have and have not met sustainability objectives as evidence of an ongoing commitment to achieving those benchmarks (Bendell & Thomas, 2013; Boston Consulting Group, 2013; Carcano, 2013; Ertekin and Atik 2015; Kapferer & Michaut, 2015). LVMH does this in its annual environmental reports as a matter of policy, but could communicate across consumer channels to reaffirm its commitment to its Environmental Charter and annual objectives as part of the "process, progress, and environmental impact" stage (see Fig. 3.1). Research suggests that companies should reinforce their legitimacy (i.e., brand image, vision, company leadership, market leadership) to add value to SLBC (Gardetti, 2016; Guercini & Ranfagni, 2013). By extending their legitimacy to the brand communication, luxury companies can facilitate the legitimation (i.e., mainstream acceptance) of sustainability and luxury as intrinsic concepts (Ertekin & Atik, 2015; Hennigs et al., 2013; Hopkinson & Cronin, 2015; Kapferer & Michaut, 2015; Rahman & Yadlapalli, 2015).

Authenticity is important to consider when developing content and messaging for SLBC (Bendell and Kleanthous, 2007; Beverland, 2006; Boston Consulting Group, 2013; Grail Research 2010; Ivan et al., 2016; Ranfagni & Guercini, 2016). It is common in luxury marketing for fictional narratives to be employed such as those that create an artificial heritage around newer luxury brands (Freire, 2014). LVMH-owned Thomas Pink is positioned as a traditional British shirt maker similar to those founded on London's Jermyn Street in the eighteenth century. By incorporating "Jermyn Street" into its brand logo and crafting a fictional origin story, Thomas Pink created an artificial heritage when, in actuality, the company was only founded in 1984. While fictional narratives typically fuel the dream value for luxury brands, this research argues that luxury brands' communication on sustainability initiatives should focus on true stories, not fiction. However, the dream value of luxury can still be activated through experiential content.

Experiential Orientation

The framework does not explicate the mix of channels (e.g., Web site, mobile app, social) that a brand should employ in the SLBC strategy, but does suggest that messages be experiential, aimed first at the educational realm and leverage additional realms (i.e., entertainment, escapist, aesthetic) where possible (Atwal and Williams, 2009). Innovative communication strategies should also be considered. For example, Fendi's 2011 campaign "fato a mano for the future" (i.e., made by hand for the future) was a series of in-store events featuring artists, sculptors, and Fendi artisans creating products from would-be discarded materials from the Fendi production process. The in-store series was educational, entertaining, and aesthetically beautiful. As stores are becoming more like entertainment spaces, staging meaningful interactions with consumers in stores can be a viable SLBC strategy.

Sustainable Luxury Brand Communication Hierarchy

Preservation and Conservation

The SLBC hierarchy prioritizes communication on preservation and conservation as the first type of SLBC by luxury companies. This type of communication aligns with the financial and functional sustainable luxury values that highlight the production integrity, quality, scarcity, and durability of luxury (Hennigs et al., 2013). Despite inconclusive findings on consumer perceptions of sustainable luxury, research suggests that consumers do perceive the luxury attributes of preservation and conservation to align with sustainability, which makes this a practical point from which companies can initiate their SLBC (Kapferer & Michaut, 2014, 2015). This communication also encompasses the rare resources (e.g., pearls, diamonds, coral) that are harvested for production and the steps the company is taking to preserve these resources for the future. Environmental and animal conservation efforts related to core businesses can also be communicated at this stage in addition to community development and heritage preservation initiatives in locations where raw materials are harvested. For example, LVMH-owned Loro Piana's preservation of Vicuna and education of Vicuna farmers in Peru has saved the species from extinction, contributed to community development in Peru, and allowed Loro Piana to corner the market on this precious resource (Kapferer & Michaut, 2014; Safe, 2017).

Process, Progress, and Environmental Impact

At this communication stage, the supply chain can be illuminated within the broader environmental context, including communication about the major environmental impact of apparel production (e.g., CO_2 emissions, water and energy consumption; Cervellon, 2013). This allows for communication on how the company minimizes environmental impact in its supply chain (i.e., sourcing, creating, manufacturing, logistics, distribution) by being vertically integrated and may transfer agency to the consumer related to sustainability at the end of the product lifecycle (e.g., servicing, waste, recycling; Gardetti, 2016). For example, consumers can be made aware that Louis Vuitton offers repair services for its leather goods, that both Louis Vuitton and Parfums Christian Dior use refillable fragrance bottles, and that Loewe offers an environmentally friendly leather cleaning service ("Environmental Report," 2017). Progress toward meeting environmental objectives can also be shared at this stage. Traceability and transparency of materials and processes throughout the supply chain can also be emphasized in addition to innovations within the supply chain that are allowing the company to minimize its environmental impact. For example, Loewe recently took part in the design of a tool for assessing the environmental impact of leather, and in 2013, LVMH partnered with an environmental research firm to develop a system of traceability for python skins using DNA (Gardetti, 2016). This helps prevent poaching by ensuring that all python skins are ethically sourced and harvested ("Environmental Report," 2017). This stage can also be used to emphasize companies' commitment to eco-design of stores, factories and office spaces (Grail Research, 2010). Content highlighting the specialized skills and training that artisans receive to continue the tradition of fine product craftsmanship concludes this stage and segues to communication about the production location. For example, in 2011, LVMH started Les Journées Particulières, a bi-annual event that opens select brands' factories and workshops to the public for two days, allowing consumers a first-hand glimpse at houses' sustainable production processes (Kapferer & Michaut, 2015). Although a few LVMH brands (e.g., Tag Heuer, Emilio Pucci, Möet and Chandon) have shared content on social media related to the event, only the LVMH parent Web site and social media accounts demonstrate a focused communication effort about the event and its purpose (Kapferer & Michaut, 2015).

Production Location

The next stage of communication emphasizes the country of origin (COO) and how true luxury brands do not delocalize production from their home countries (Kapferer, 2015). Maintaining the legacy of local craftsmanship and savoir-faire can be communicated in addition to information about how the company gives back to its home country. For example, many luxury brands are investing in cultural restoration projects in their home countries when governments cannot afford to subsidize such ventures (Cervellon, 2013). In 2016, Fendi's Trevi Fountain restoration and Bulgari's Spanish Steps restoration in Rome were both completed, signifying the LVMH-owned Italian brands' commitment to their country of origin ("Environmental Report," 2017). Patronage of the arts and culture within the COO can also be communicated at this stage such as the development of the Frank Gehry-designed Louis Vuitton museum in Paris, a creative vision of LVMH CEO-Bernard Arnault ("Environmental Report," 2017). In line with research findings on sustainability marketing, sponsorship of these types of initiatives by LVMH brands has been underscored by a sense of national pride and devoid of commercial motives (Kapferer & Valette-Florence, 2016).

Partnerships, Collaboration, and Innovation

The final communication stage emphasizes sharing the company's involvement in SD and CSR initiatives that are an extension of company operations. Philanthropic and cause-related initiatives (e.g., Louis Vuitton + Unicef) have been placed in this communication stage due to the potential for consumers to perceive these initiatives as merely "greenwashing" (Cervellon, 2013). The researcher proposes that increased brand communication on other types of initiatives (e.g., preservation and conservation) by luxury brands will lend credibility to philanthropic and cause-related communication as consumers gain a broader understanding of the company's SD and CSR commitments (Guercini & Ranfagni, 2013). This stage also includes communication about the company's industry partnerships geared toward common sustainability goals, recent innovations made through R&D that will further its sustainability agenda, and collaborations with entities outside of the retail sector (e.g., governments, NGOs; Carcano, 2013).

For example, LVMH participates in the annual European Sustainable Development Week to stay abreast of new developments, challenges, and opportunities in the field of sustainable development. LVMH also became the first luxury organization to impose a voluntary carbon tax in 2015, announcing its implementation at the 21st United Nations Climate Change Summit in Paris in order to demonstrate its commitment to environmental policy and to encourage like-minded organizations to follow suit in implementing their own carbon taxes (Winston, 2017). LVMH's star brand, Louis Vuitton, was a main sponsor of the Summit which further demonstrated the conglomerate's commitment to organizational leadership in SD.

Conclusions and Recommendations

This conceptual study explored the need for experiential SLBC as a means of knowledge transfer with the aim of educating and increasing awareness of luxury brands' SD and CSR initiatives among luxury consumers, and perhaps, the wider consumer market and apparel industry. Communicating on SD and CSR initiatives and the steps being taken to minimize environmental impact may also create an opportunity for luxury companies to spread awareness about larger societal and environmental issues (e.g., global warming, climate change). Progressive luxury companies like LVMH, whose long-standing sustainability agenda is rooted in organizational core values and a forward-thinking strategic approach, can leverage their brand power as trendsetters and opinions leaders more broadly to initiate the dissemination of sustainability knowledge through increased brand communication, a necessary step to fuel the shift to a more sustainable global retail industry and a culture of sustainable consumption.

This research, although necessarily introductory in its findings, contributes to the growing body of research on sustainable luxury and is the first to offer a theoretical conceptualization of SLBC. The framework is intended to inform the development of communication strategies by luxury practitioners and serve as a guide for future academic study. To that end, there are many directions for future research. Luxury consumers' perceptions of and attitudes toward companies' SLBC strategies should be empirically measured in addition to changes in consumers' levels of awareness and knowledge that occur as a result of exposure to and engagement with the brand communication. These explorations will significantly strengthen this research area and may provide meaningful managerial implications.

References

Achabou, M. A., & Dekhili, S. (2013). Luxury and sustainable development: Is there a match? *Journal of Business Research, 66*(10), 1896–1903.

Atwal, G., & Williams, A. (2009). Luxury brand marketing: The experience is everything. *Journal of Brand Management, 16*(5/6), 338–346.

Beverland, M. (2006). The 'real thing': Branding authenticity in the luxury wine trade. *Journal of Business Research, 59*(2), 251–258.

Bendell, J., & Kleanthous, A. (2007). *Deeper luxury*. London: WWF.

Bendell, J., & Thomas, L. (2013). The appearance of elegant disruption: Theorising sustainable luxury entrepreneurship. *Journal of Corporate Citizenship, 52*, 9–25.

Berthon, P., Pitt, L., Parent, M., & Berthon, J. P. (2009). Aesthetics and ephemerality: Observing and preserving the luxury brand. *California Management Review, 52*(1), 45–66.
Boston Consulting Group. (2013). *How millennials are changing the face of marketing forever*. Retrieved from https://www.bcgperspectives.com/content/articles/marketing_center_consumer_customer_insight_how_millennials_changing_marketing_forever/#chapter1
Carcano, L. (2013). Strategic management and sustainability in luxury companies. *Sustainable Luxury: A special theme issue of The Journal of Corporate Citizenship, 52,* 36–54.
Cavender, R., & Kincade, D. H. (2015). A luxury brand management framework built from historical review and case study analysis. *International Journal of Retail & Distribution Management, 43*(10/11), 1083–1100.
Cervellon, M. C. (2013). Conspicuous conservation: Using semiotics to understand sustainable luxury. *International Journal of Market Research, 55*(5), 695–717.
Closs, D. J., Speier, C., & Meacham, N. (2011). Sustainability to support end-to-end value chains: the role of supply chain management. *Journal of the Academy of Marketing Science, 39*(1), 101–116.
Dach, L., & Allmendinger, K. (2014). Sustainability in corporate communications and its influence on consumer awareness and perceptions: A study of H&M and Primark. *Procedia-Social and Behavioral Sciences, 130,* 409–418.
Davies, I. A., Lee, Z., & Ahonkhai, I. (2012). Do consumers care about ethical-luxury? *Journal of Business Ethics, 106*(1), 37–51.
Environmental Report 2016. (2017). *LVMH*. Retrieved from www.lvmh.com
Ertekin, Z. O., & Atik, D. (2015). Sustainable markets motivating factors, barriers, and remedies for mobilization of slow fashion. *Journal of Macromarketing, 35*(1), 53–69.
Faust, M. E. (2013). Cashmere: A lux-story supply chain told by retailers to build a competitive sustainable advantage. *International Journal of Retail & Distribution Management, 41*(11–12), 973–985.
Fionda, A. M., & Moore, C. M. (2009). The anatomy of the luxury fashion brand. *Journal of Brand Management, 16*(5–6), 347–363.
Freire, N. A. (2014). When luxury advertising adds the identitary values of luxury: A semiotic analysis. *Journal of Business Research, 67*(12), 2666–2675.
Gardetti, M. A. (2016). Loewe: Luxury and sustainable management. *Handbook of Sustainable Luxury Textiles and Fashion, 2,* 1–16.
Gardetti, M. A., & Giron, M. E. (Eds.). (2014). *Sustainable luxury and social entrepreneurship: Stories from the Pioneers*. Shipley, United Kingdom: Greenleaf Publishing.
Grail Research. (2010). *Green—The new color of Luxury: Moving to a sustainable future*. Retrieved from www.grailresearch.com/pdf/ContenPodsPdf/2010-Dec-Grail-Research-Green-The-New-Color-of-Luxury.pdf
Guercini, S., & Ranfagni, S. (2013). Sustainability and luxury. The Italian case of a supply chain based on native wool. *Journal of Corporate Citizenship, 52,* 76–89.
Hennigs, N., Wiedmann, K. P., Klarmann, C., & Behrens, S. (2013). Sustainability as part of the luxury essence: Delivering value through social and environmental excellence. *Journal of Corporate Citizenship, 52,* 25–35.
Hopkinson, G. C., & Cronin, J. (2015). When people take action…. Mainstreaming malcontent and the role of the celebrity institutional entrepreneur. *Journal of Marketing Management, 31* (13–14), 1383–1402.
Ivan, C. M., Mukta, R., Sudeep, C., & Burak, C. (2016). Long-term sustainable sustainability in luxury. Where else? In *Handbook of Sustainable Luxury Textiles and Fashion* (pp. 17–34). Springer Singapore.
Janssen, C., Vanhamme, J., Lindgreen, A., & Lefebvre, C. (2014). The Catch-22 of responsible luxury: Effects of luxury product characteristics on consumers' perception of fit with corporate social responsibility. *Journal of Business Ethics, 119*(1), 45–57.
Joy, A., Sherry, J. F., Venkatesh, A., Wang, J., & Chan, R. (2012). Fast fashion, sustainability, and the ethical appeal of luxury brands. *Fashion Theory, 16*(3), 273–295.

Joy, A. (2015, June 22). Fast fashion, luxury brands, and sustainability. *The European Financial Review*. Retrieved from http://www.europeanfinancialreview.com/?p=4589.

Kaperer, J. N. (2010). All that glitters is not green: The challenge of sustainable luxury. *European Business Review*, November–December, 40–45.

Kapferer, J. N. (2015). *Kapferer on luxury: How luxury brands can grow yet remain rare*. Philadelphia, PA: Kogan Page Publishers.

Kapferer, J. N., & Bastien, V. (2009). The specificity of luxury management: Turning marketing upside down. *Journal of Brand Management, 16*(5), 311–322.

Kapferer, J. N., & Michaut, A. (2014). Is luxury compatible with sustainability? Luxury consumers' viewpoint. *Journal of Brand Management, 21*(1), 1–22.

Kapferer, J. N., & Michaut, A. (2015). Luxury and sustainability: a common future? The match depends on how consumers define luxury. *Luxury Research Journal, 1*(1), 3–17.

Kapferer, J. N., & Valette-Florence, P. (2016). Beyond rarity: The paths of luxury desire. How luxury brands grow yet remain desirable. *Journal of Product & Brand Management, 25*(2), 120–133.

LaRocca, D. (2014). Brunello Cucinelli: A humanistic approach to luxury, philanthropy, and stewardship. *Journal of Religion and Business Ethics, 3*(1), 1–26.

Lovegrove, D. (2011). For richer or poorer? For the sake of planet Earth. *Raconteur on Sustainable Luxury, 7*. Retrieved from http://np.netpublicator.com/netpublication/n10444899

Muratovski, G. (2015). Sustainable consumption: Luxury branding as a catalyst for social change. In M. A. Gardetti & A. L. Torres (Eds.). *Sustainable luxury*. Shipley, United Kingdom: Greenleaf Publishing.

Phau, I., & Prendergast, G. (2000). Consuming luxury brands: The relevance of the 'rarity principle'. *Journal of Brand Management, 8*(2), 122–138.

Pine, B. J., & Gilmore, J. H. (1998). Welcome to the experience economy. *Harvard Business Review, 76*(4), 97–105.

Rahman, S., & Yadlapalli, A. (2015). Sustainable practices in luxury apparel industry. *Handbook of Sustainable Luxury Textiles and Apparel, 1,* 187–211.

Ranfagni, S., & Guercini, S. (2016). Beyond appearances: The hidden meanings of sustainable luxury. *Handbook of Sustainable Luxury Textiles and Fashion, 2,* 51–72.

Safe, G. (2017, June 17). Luxury brands are snapping up farms to control their supply chains. *Business of Fashion*. Retrieved from https://www.businessoffashion.com.

Schmitt, B. (1999). Experiential marketing. *Journal of Marketing Management, 15*(1–3), 53–67.

Solomon, M. (2017, June 20). How Millennials will reshape the luxury market. *Forbes*. Retrieved from https://www.forbes.com.

Vigneron, F., & Johnson, L. W. (2004). Measuring perceptions of brand luxury. *Journal of Brand Management, 11*(6), 484–506.

Winston, A. (2017, January 11). An inside view of how LVMH makes luxury more sustainable. *Harvard Business Review*. Retrieved from https://hbr.org/2017/01/an-inside-view-of-how-lvmh-makes-luxury-more-sustainable.

Chapter 4
Luxury Fashion Brands Versus Mass Fashion Brands: Data Mining Analysis of Social Media Responses Toward Corporate Sustainability

Stacy Hyun-Nam Lee, Yi Zhou, Chris K. Y. Lo and Jung Ha-Brookshire

Abstract Today's consumers are increasingly concerned with social and environmental issues, leading to more conscientious consumption decisions and practices. As the core consumers' characteristics are vastly different between luxury and mass fashion brands, it is expected that consumers' social media responses would be highly varied as well. Therefore, this study aimed to explore consumer social media behavior when exposed to corporate sustainability messages and discover potential differences in responses between luxury and mass fashion consumers. To achieve the objective of the study, 89,290 raw data were obtained from Twitter through Python. Given that there might be differences in consumer responses toward corporate 378 sustainability messages between luxury and mass fashion brands because of distinctively different characteristics in their target consumers, this study explored 380 consumer social media behavior and looked for potential differences in responses 381 between the two groups of consumers. After analyzing over 89,000 tweets and 382 retweets made by 11 luxury and 11 mass fashion brands as of March 2017, the 383 study found several interesting results. Overall, the analysis of Twitter messages suggests that luxury fashion brands are less communicative with consumers about their sustainability activities than mass fashion brands. This indicates fewer tweet and less loud, yet effective in what they

S. H.-N. Lee (✉) · Y. Zhou · C. K. Y. Lo
Business Division, Institute of Textiles and Clothing, The Hong Kong Polytechnic University, Hung Hom, Kowloon, Hong Kong
e-mail: Stacy.hn.lee@polyu.edu.hk

Y. Zhou
e-mail: yi-paul.zhou@connect.polyu.hk

C. K. Y. Lo
e-mail: kwan.yu.lo@polyu.edu.hk

J. Ha-Brookshire
Textile and Apparel Management, College of Human Environmental Sciences, University of Missouri, 137 Stanley Hall, Columbia, MO 65211, USA
e-mail: habrookshirej@missouri.edu

© Springer Nature Singapore Pte Ltd. 2018
C. K. Y. Lo and J. Ha-Brookshire (eds.), *Sustainability in Luxury Fashion Business*, Springer Series in Fashion Business, https://doi.org/10.1007/978-981-10-8878-0_4

communicate might be luxury brands' strategies. Indeed, the findings do show that consumers look for leadership in luxury brands, as they tend to like or retweet more messages generated from luxury brands when their messages are focused on sustainability.

Today's consumers are increasingly concerned with social and environmental issues, leading to more conscientious consumption decisions and practices (Hennings, Widmann, Klarmann, & Behrens, 2013). This consumer behavior also involves a greater demand for more transparent information about brands and products, not only regarding the positive aspects of brands, but also for the controversial issues related to brands' social and environmental impact (De Vries, Gensler, & Leeflang, 2012). Therefore, many leading brands have been pressured to integrate sustainability goals into their business strategies, and more brands are taking further steps toward achieving their sustainability goals.

Despite fashion brands' increasing push toward corporate sustainability, the strategies for communicating their sustainability efforts have been inconsistent (DEFRA, 2011; Goworek, Fisher, Cooper, Woodward, & Hiller, 2012; Yan, Hyllegard, & Blaesi, 2012). The economic dimension of sustainability is clearly relevant to the profit-driven nature of businesses, and positive results can result from producing or delivering efficiently in this area (Dao, Langella, & Carbo, 2011). Therefore, it is imperative to communicate brands' initiatives and achievements toward achieving sustainability. However, reporting and integrating the social and 26 environmental aspects of sustainability is challenging due to the intangible nature of 27 these goals (Henriques & Richardson, 2013). Therefore, the absence of clear communication from brands regarding their sustainability practices often results in consumers being unaware of brands' sustainability efforts across categories, thereby preventing consumers from purchasing sustainable products (Finnerty, Stanely, & Herther, 2013).

Furthermore, as compared to mass fashion brands, consumers expect luxury brands to tackle sustainability to a greater degree and, at the same time, demand responses to inquiries about environmental and social issues (Bendell & Kleanthous, 2007). Besides, more consumers want to obtain transparent information about the sourcing and manufacturing of products (Bhaduri & Ha-Brookshire, 2011). To respond to such demands, luxury brands, such as Stella McCartney, Ferragamo, and Vivienne Westwood, started to incorporate sustainability as a key marketing feature (Davies, Lee, & Ahonkhai, 2012). At the same time, mass fashion brands, such as H&M and Adidas, are also investing heavily in sustainability efforts. While the previously mentioned luxury brands tend to exercise sustainability in a philosophical or sometimes even emotional level, mass fashion brands focus on innovations in recycling materials or reducing waste in their product and production design.

Both luxury and mass fashion brands communicate their sustainability efforts via social media, attempting to reach out to their consumers and stakeholders to convince them of their efforts toward achieving sustainability goals. Given that the core consumers' characteristics are vastly different between luxury and mass fashion brands, it is expected that consumers' social media responses would be highly varied as well. However, we know little about these differences in response patterns. Therefore, this study aimed to explore consumer social media behavior when exposed to corporate sustainability messages and discover potential differences in responses between luxury and mass fashion consumers. The next section discusses the literature on sustainability in both luxury and mass fashion brands, followed by methods, results, and conclusions.

Literature Review

Sustainability and Luxury Fashion Brands

Luxury fashion brands offer uniqueness, value, and exclusivity; customers acquire goods for what they symbolize and signify (Belk, 2014). As luxury fashion brands often represent consumer identity and individual value system, responsible and conscientious consumers expect luxury fashion brands' sustainability values to reflect their own, and thus demand responses to inquiries of corporate sustainability (Bendell & Kleanthous, 2007). Accordingly, numerous luxury fashion brands, such as Stella McCartney, Ferragamo, and Vivienne Westwood, started to implement sustainability as a part of the luxury brands' essence (Davies et al., 2012; Joy, Sherry, Venkatesh, Wang, & Chan, 2012).

While balancing brand exclusivity and the ubiquitous nature of the brands' globalization, luxury fashion brands often restructure their business models to combat a variety of challenges, such as the wide availability of counterfeits, poor labor standards and sweatshops, blood diamonds, irresponsible gold-mining practices, and even anorexic models (Godey et al., 2016; Janssen, Vanhamme, Lindgreen, & Lefebvre, 2014). For instance, the luxury cosmetics firm Garnier was found guilty of racial discrimination, and LVMH, owners of Louis Vuitton and TAG Heuer, were removed from the FTSE4Good index due to poor compliance with supply chain requirements. Prada has also been criticized for exploiting illegal Chinese immigrant workforces in Italian sweatshops (Joy et al., 2012). Particularly in emerging markets, luxury brands have been accused of exacerbating social disparities and the tensions between the rich and the poor (Kleanthous, 2011).

Such scandals stemming from unsustainable business practices could be detrimental to luxury fashion brands because of: (a) their global presence and access to a tremendous amount of people worldwide and (b) the leadership characteristics that their core consumers possess within social classes. Therefore, researchers point out that luxury fashion brands must prioritize corporate sustainability in coming years because such negative reports could greatly harm their reputation (Janssen et al., 2014; Kleanthous, 2011).

Sustainability and Mass Fashion Brands

In contrast to luxury fashion brands, today's mass fashion marketplace is filled with fast fashion brands that are known as "inexpensive clothing that is meant to be worn a limited number of times" (Fulton & Lee, 2013). Fundamentally, to increase profits for fast fashion brands, the shorter production and distribution lead times have become important for fast fashion brands' business model, which in turn amplifies the availability of highly time-sensitive clothing in the marketplace (Cachon & Swinney, 2011). To fulfill consumers' demand for fast fashion, brands tend to use lower quality materials and manufacturing processes, which, in turn, accelerates the obsolescence of the products (Kozlowski, Searcy, & Bardecki, 2014). Eventually, the disposability of fashion products encouraged by the fast fashion business model has changed consumer consumption patterns (Allwood, Laursen, Malvido de Rodríguez, & Bocken, 2006; Fletcher, 2012). Sustainability researchers point out the very nature of fast fashion and its unnecessary textile waste and negative environmental impact throughout the entire value chain by referring to it as "McFashion" (Joy et al., 2012; Kozlowski et al., 2014).

To combat such criticism, fast fashion brands are becoming more active in practicing corporate sustainability activities because the demand for sustainable products is increasing especially among the Millennials, the core consumers of fast fashion products (Hulm & Domeisen, 2008; Li, Zhao, Shi, & Li, 2014). For instance, global fast fashion brands, such as ZARA, H&M, Gap, and UNIQLO, are exercising green marketing strategies to affect consumers' purchasing behaviors as well as to lead suppliers into strategic alliances (Li et al., 2014). In order to decrease the toll on the environment, other brands, such as Timberland and Levi Strauss, have changed their production process to become more environmentally friendly by using recycled materials and eliminating harmful chemicals (Park & Lennon, 2006). Other brands, such as Nike and Gap, have also begun to blend organic cotton into their cotton product supply chain (Atasu, 2016; Bonner, 1997).

Sustainability and Social Media Brand Communication

With the advent of social media, many companies and brands have faced a new era in creating interactive ways to engage their customers (Gallaugher & Ransbotham, 2010). Social media has changed the way brand content is created, distributed, and consumed, and this trend has fundamentally shifted the power of brand image management from brand marketers to consumers (Tsai & Men, 2013). As a marketing tool, numerous brands are increasingly investing their marketing efforts toward digital communication and have become pioneers in the ways they utilize various social media platforms (Heine & Berghaus, 2014). Particularly, global luxury brands have been able to build brand reputation by reaching out to more than two-thirds of all Internet users in the world (Correa, Hinsley, & De Zúñiga, 2010).

Undoubtedly, advances in technology and digital communications, coupled with a changing global economy, have had a significant influence on consumer perceptions and experiences with global brands (Dhaoui, 2014).

Overall, more than 50% of social media users follow brands on social media (De Vries et al., 2012). In addition, about US$4.3 billion is being spent for worldwide marketing on social networking sites (De Vries et al., 2012). This has intensified competition between brands on social media and has further evolved into user-generated branding (UGB). UGB co-owns brands, co-directs brands' competitive strategies, and co-defines symbolic meanings of the brands with consumers (Heil, Lehmann, & Stremersch, 2010). In this way, consumers become more knowledgeable about brands and are better able to make consumption decisions through business-to-consumer interactions on various online platforms. Certainly, advances in technology and digital communications, coupled with a changing global economy, have had a significant influence on consumer perceptions and experiences with brands (Dhaoui, 2014).

Although social media platforms help strengthen many fashion brands' digital presence, electronic word-of-mouth (e-WOM) communication can be a double-edged sword, as consumers share and retweet not only positive opinions but also negative opinions that could change others' perceptions of the brands. Therefore, fashion brand managers must pay attention not only to positive but also negative consumer feedback as this information could be shared by thousands or even millions over a very short time period. When consumers are dissatisfied with certain products or services, they may be more motivated to express their experiences in the form of a complaint, directly on a social media site (Einwiller & Steilen, 2015). Prospective customers will choose to refrain from interacting with the given company as they may perceive a higher level of risk and consider the company to be untrustworthy (Kim, Ferrin, & Rao, 2008). Positive reputations and responses are integral for reducing the possibility of risk or negative information within the context of purchasing products and services, either online or offline, as e-WOM is strongly associated with risk perceptions and consumers' trust in a given company (Kim et al., 2008).

Therefore, the literature suggests that corporate sustainability communication must also consider the e-WOM phenomenon. Because of the uniquely different characteristics of core consumers between luxury fashion and mass fashion brands, the behavior of social media followers of luxury and mass fashion brands may differ and thus vary their responses to corporate sustainability communication presented on social media platforms. However, little research exists comparing consumer social media responses to corporate sustainability communication between luxury and mass fashion brand consumers. To fill this gap, the purpose of this research was to explore consumer social media responses to corporate sustainability communication, and gauge potential differences in responses between the two groups of consumers.

Method and Analysis

Research Design

For the purposes of the study, we conducted a systematic content analysis of the corporate sustainability messages that were posted by luxury and mass fashion brands. Twitter was the primary social media platform used in this study because it has moved ahead of Facebook and Google+ to become the fastest-growing social media outlet where users can share (or tweet) their thoughts, links, and pictures (Bennett, 2013). Since its launch in 2006, over 75% of the Fortune Global 100 companies possess one or more Twitter accounts at the corporate level as well as separate additional accounts for their specific brands (Malhotra, Malhotra, & See, 2012). By following and being followed by others, three types of tweets can be created: original tweets, replies, and retweets, and thus information can be quickly disseminated (Chae, 2015). Specifically, original tweets can be created on the sender's profile page and home timeline, which can be retweeted by other users, or users can join conversations by @replying to others. As Twitter offers Application Programming Interface (API) and is traceable, numerous researchers, practitioners, and organizations have been vastly captivated with collecting and analyzing tweets by using diverse types of queries, such as keywords and user profiles (Twitter, 2013). Consequently, Twitter data have been used in abundant practical applications, from engaging with real-time events to trend analysis (Dickey, 2014), brand management (Malhotra et al., 2012), and crisis management (Wyatt, 2013).

Given that 500 million tweets are created per day by over 270 active Twitter users, Twitter provides "open" data, offering research and business opportunities to access the data on an unprecedented scale and size, using the Twitter Application Programming Interface (API) (Twitter, 2013). Typically, keywords and hashtags can be searched and downloaded through the programming tool "Python" for Twitter API to obtain and mine data. Thus, previous research on Twitter has aimed to analyze data for challenging problems in diverse domains, including norms and behaviors, self-presentation (Marwick & Boyd, 2011) and to explore why people post (Jansen, Zhang, Sobel, & Chowdury, 2009). Particularly because of Twitter's accessible APIs, inherent openness, and rampant worldwide popularity, it is an ideal social Web site to zoom in on and find relevant patterns of trending topics related to socialization (Sadat, Ahmed, & Mohiuddin, 2014). Accordingly, researchers argue that Twitter brings great opportunities for exploring different methodologies from various intellectual backgrounds, including descriptive analysis and content analysis (Chae, 2015).

Data Collection

To achieve our study's objective, we reviewed all tweets and their related information posted by the top 11 fashion luxury brands (i.e., Armani, Burberry, Cartier,

Chanel, Coach, Dior, Fendi, Gucci, Hermes, Louis Vuitton, and Prada) and 11 mass fashion and sportswear brands (i.e., Adidas, C&A, Gap, H&M, Levis, Mango, Marks & Spencer, Nike, Puma, Topshop, and Zara). We used these 22 brands due to their substantial role in the fashion industry, dominating the market in their respective categories. We used the programming tool "Python v3.6.1" to develop a program to use Twitter's REST APIs, which provide programmatic access to read and write Twitter data.[1] By using the unique Twitter account names for the above 22 brands, the program called Twitter Search API (https://twitter.com/search-advanced) downloaded all tweets posted by each brand and their metadata (e.g., each tweet's ID, creation date, numbers of likes, numbers of retweets, hashtags). The data were collected starting from the creation date of each brand's Twitter account to March 31, 2017. Thus, we collected a total of 342,473 tweets and metadata (including general tweets, replies, and retweets). We deleted 253,183 replies because they were not initially posted by the brands themselves and thus cannot directly express the brands' attitude toward sustainability issues. After deleting these replies, we prepared 89,290 general tweets and retweets for further analysis.

Keyword Search and Analysis

We conducted keyword searches to identify the tweets that related to sustainable issues. We first created a keyword list based on (a) the sustainability keyword list developed by University of North Carolina sustainability office and (b) the keywords from Jung and Ha-Brookshire's (2017) study (all keywords are shown in Appendix). Second, we searched the 89,290 tweets for the keywords (and their different word formations) in our keyword list. If a tweet contained at least one of these keywords, we identified it as a sustainability tweet. After this identification, we double-checked the tweet's content to ensure it was related to sustainability issues. By using the keyword search, we found 711 sustainability tweets (i.e., 602 tweets from mass fashion brands and 109 tweets from luxury brands).

We collected and analyzed the data to learn how corporate sustainability messages with one or several of these keywords were being "liked" or "retweeted," which is the most direct way to measure whether such messages are well-received by social media users (da Silva Zago & Bastos, 2013; Riguelme & González-Cantergiani, 2016). We also looked for the keywords' performance by brand and the combinations of keywords in all the tweets' available records. Finally, we investigated why some tweets received much greater attention compared to the norm, in order to improve brands' communication about sustainability issues on social media.

[1]Please refer to Twitter Developer Documentation for REST APIs (https://dev.twitter.com/rest/public) for the detail functions.

Results

Descriptive Analysis

Descriptive analysis centers around descriptive statistics, such as the number of tweets, distribution of different types of tweets, and number of hashtags (Chae, 2015). By extracting data through examining the number of tweets and hashtags, we obtained detailed information on who tweets, replies, and retweets, which can be closely related to the brands' popularity. To achieve the study's objective, we compared the descriptive statistics of two clusters: all 89,290 tweets and 711 sustainability tweets. The results are as follows:

Figures 1, 2, and 3 present the total numbers of tweets, liked, and retweeted for all tweets and sustainability tweets categorized by mass fashion brands and luxury

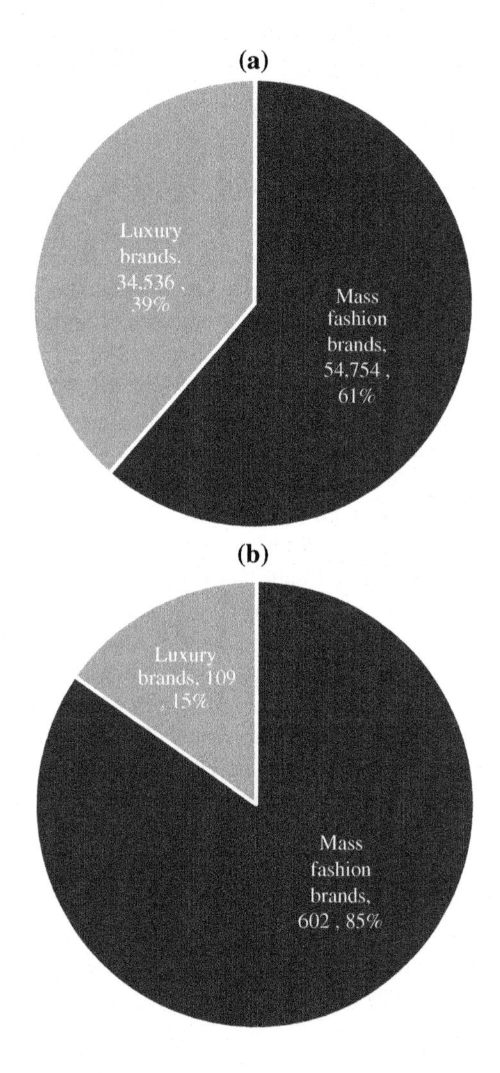

Fig. 1 **a** Total number of tweets (all tweets), 89,290/ 100%. **b** Total number of tweets (sustainability tweets), 711, 100%

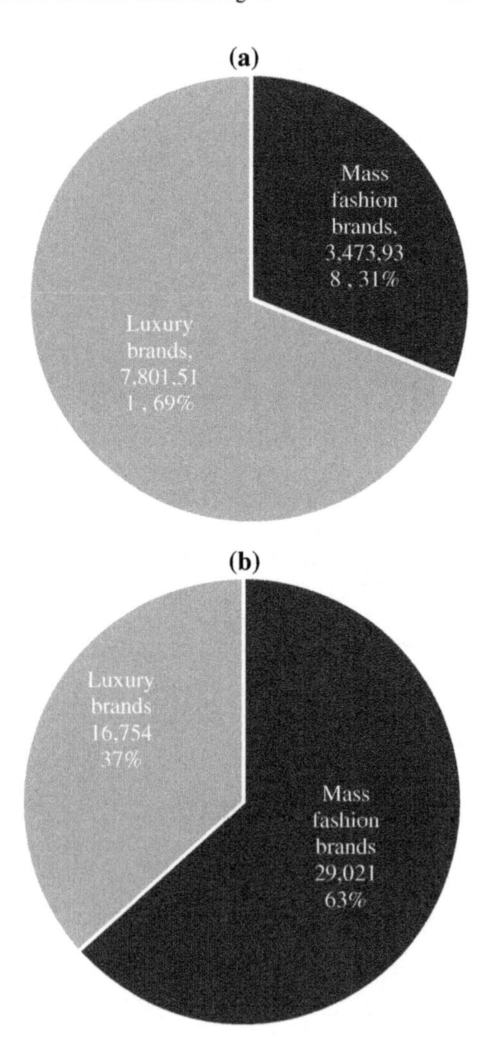

Fig. 2 **a** Total number of likes (all tweets), 11,275,449/100%. **b** Total number of likes (sustainability tweets), 45,775/100%

brands. The sizes of the pie chart in the figures picture the number of tweets liked and retweeted. As you can see from the figures, the size of the "All tweets" area is more than 100 times larger than that of the "Sustainability tweets" area.

According to Fig. 1a, the mass fashion brands posted 54,754 out of 89,290 total tweets (61% of all tweets), while the luxury brands posted 34,536 out of 89,290 tweets (39% of all tweets). Among the 711 sustainability tweets (Fig. 1b), 85% were from mass fashion brands and the other 15% were from luxury brands. This indicates that mass fashion brands post more tweets than luxury brands in either case. Out of those tweets, only 0.3% (109 out of 34,536 tweets) were related to sustainability for the luxury brands, and 1.1% (609 out of 54,754 tweets) for the mass fashion brands. This indicates that mass fashion brands more often discussed sustainability on Twitter than did luxury brands. This emphasizes an important

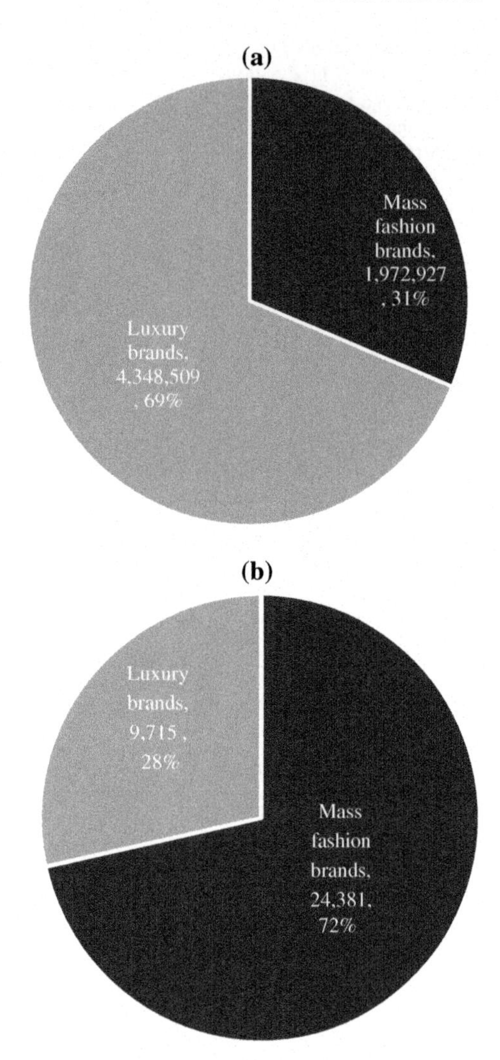

Fig. 3 **a** Total number of retweets (all tweets), 6,321,436/100%. **b** Total number of retweets (sustainability tweets), 34,140/100%

result: That mass fashion brands are far more engaged in consumers' perceptions about sustainability than luxury brands. More specifically, all 11 mass fashion brands were engaged in sustainability while only 7 out of 11 luxury brands were involved.

Figures 2a and 3a show that the total number of likes and retweets for the luxury brands was comparatively higher than those for the mass fashion brands in the "All tweets" category. Overall, luxury brands received more attention from customers than mass fashion brands on Twitter. For sustainability tweets (Figs. 2b and 3b), the total number of likes (retweets) is 45,775 (34,140) for both mass fashion and luxury brands. Although tweets about mass fashion brands' total sustainability practices were liked 1.7 times (or retweeted 2.5 times) more than luxury brands, it cannot be

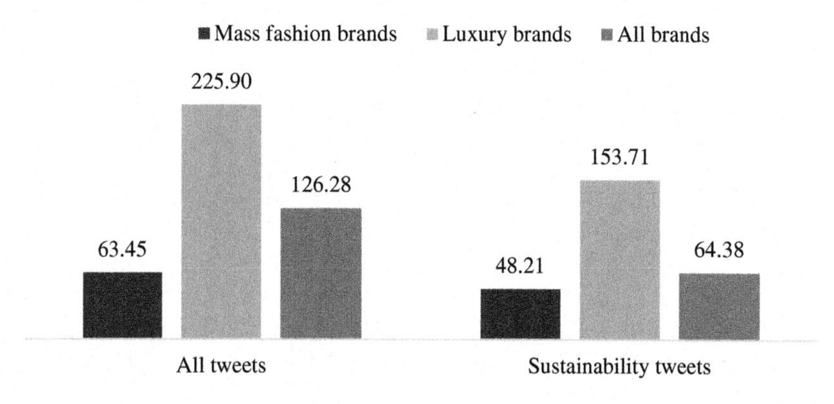

Fig. 4 Average number of likes

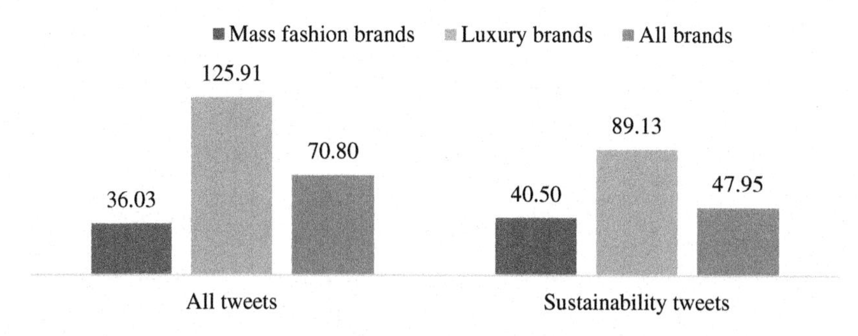

Fig. 5 Average number of retweets

presumed that the performance of mass fashion brands was better than that of luxury brands. Only 15% of sustainability tweets were posted by luxury brands, but they contributed 37% (28%) of the total number of likes (retweets). Thus, the average number of likes and retweets received per tweet for mass fashion brands and luxury brands was examined to compare the performance of the brands in Figs. 4 and 5.

According to Figs. 4 and 5, luxury brands received more likes and retweets per tweet than mass fashion brands for both "All tweets" and "Sustainability tweets." More of the luxury brands' messages were liked and retweeted, including sustainability messages, with an average frequency three times higher (or retweeted twice higher) than that of mass fashion brands. This suggests that the influence of luxury brands is far greater than mass fashion brands. More importantly, in terms of sustainability messages as compared to overall tweets, Adidas and Ca-Europe are the strongest among the mass fashion brands while the luxury brands Armani and Coach are stronger in their category. On the other hand, Adidas, Ca-Europe, M&S, Puma, Armani, and Coach have higher numbers of retweets as compared to their general tweets.

Content Analysis

Content analysis was conducted by performing a word search. First, the sustainability conversations on Twitter by the brands were identified. Multiple levels of keyword analysis (e.g., single keyword and multi-keyword combinations) related to sustainability were investigated by reviewing term frequency and comparing both luxury and mass fashion brands using level analysis (Chae, 2015). Out of 711 sustainability tweets, a total of 29 keywords in our keyword list were found, with a frequency of 29 average uses per keyword. Figure 6 shows the distribution of the top 15 most tweeted words out of the 29 keywords. The top 15 most tweeted words (usage frequency) were: "donate" (201), "charity" (123), "sustain" (119), "water" (98), "recycle" (75), "fair" (28), "earth" (22), "environment" (20), "save" (19), "green" (16), "RED" (16), "clean" (14), "waste" (14), "planet" (13), and "safe" (13).

Subsequently, the findings suggest that sustainability-related tweets attract more public attention on Twitter (i.e., numbers of likes and retweets). Figure 7 indicates the average numbers of likes and retweets for the 29 keywords. Most frequently used words usually receiving large numbers of likes and retweets can be seen on the bar chart. Yet, interestingly, some keywords were not used by the brands frequently, but still received large numbers of likes and retweets, such as "ocean," "climate," "renewable," "toxic," and "plant."

We further examined the differences in sustainability content between mass fashion brands and luxury brands (see Figs. 8 and 9). Generally, mass fashion brands used more sustainability-related keywords (28 out of 29 keywords) than

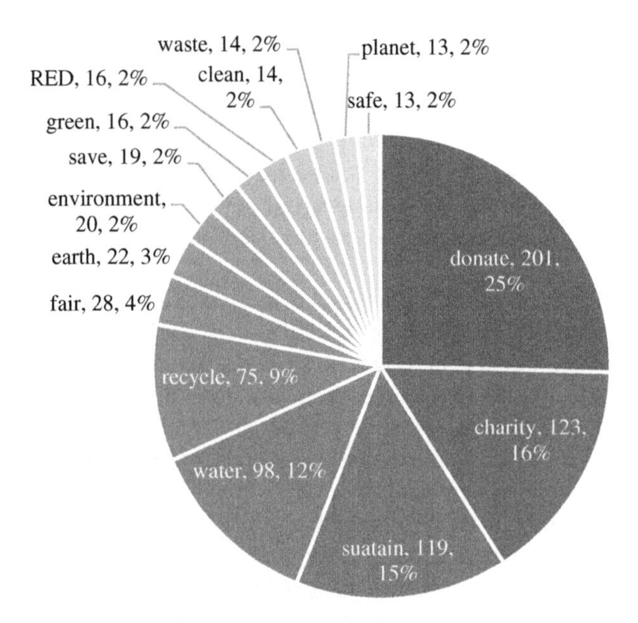

Fig. 6 Distribution of top 15 most tweeted words (all brands)

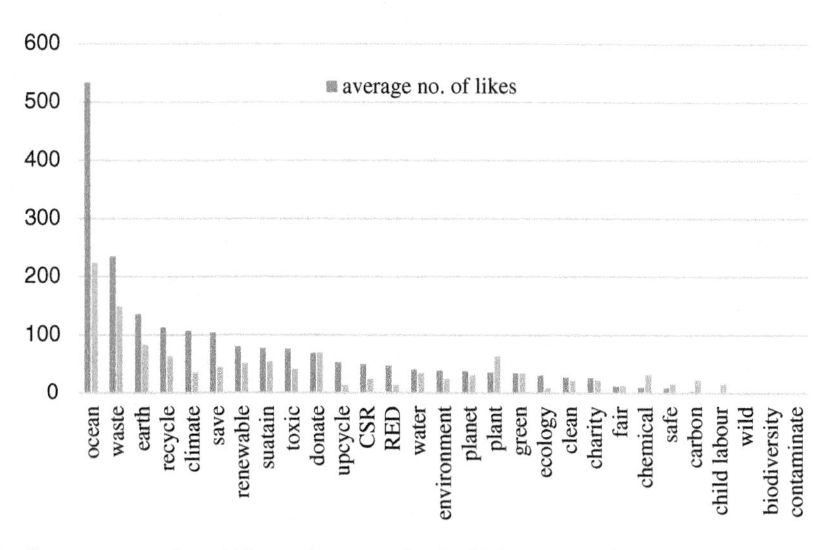

Fig. 7 Average number of likes and retweets for the 29 keywords (all brands)

luxury brands (19 out of 29 keywords). Figure 8a, b presents the keyword preferences of the mass fashion and luxury brands. As we can see from the figures, both mass fashion and luxury brands used "charity" and "donate" frequently. "Sustain" and "water" were also used more often than other words, especially for mass fashion brands. Some words, such as "recycle," "fair," "RED," "clean," "waste," and "planet," were sometimes used by mass fashion brands, but seldom by luxury brands.

Figure 9a, b indicates the average numbers of likes and retweets for the keywords used by mass fashion and luxury brands. The performance was quite different between mass fashion and luxury fashion brands. The well-performing tweets of mass fashion brands usually included the words "ocean," "waste," "save," and "recycle," while luxury brands usually included "donate," "earth," "water," and "climate."

A tweet may include multiple keywords. There were 59 multi-keyword combinations in our sample. Thus, we explored the usage and performance of these multi-keyword combinations. Figure 10 shows the distribution of frequency usage of the top 10 most tweeted multi-keyword combinations. Brands usually tweeted combinations of keywords such as "clean" and "water," "recycle" and "plant," "donate" and "charity," and "recycle" and "waste." Furthermore, Fig. 11 illustrates the average number of likes and retweets for the top 10 most liked multi-keyword combinations. Interestingly, the most liked multi-keyword combinations were not from the most tweeted combinations. For instance, the multi-keyword combination "ocean" and "save" was only used once by Adidas but received much attention (1,594 likes and 497 retweets). The "clean" and "water" combination was used by H&M and Levis 10 times, but only received an average of 22 likes (23 retweets) per usage.

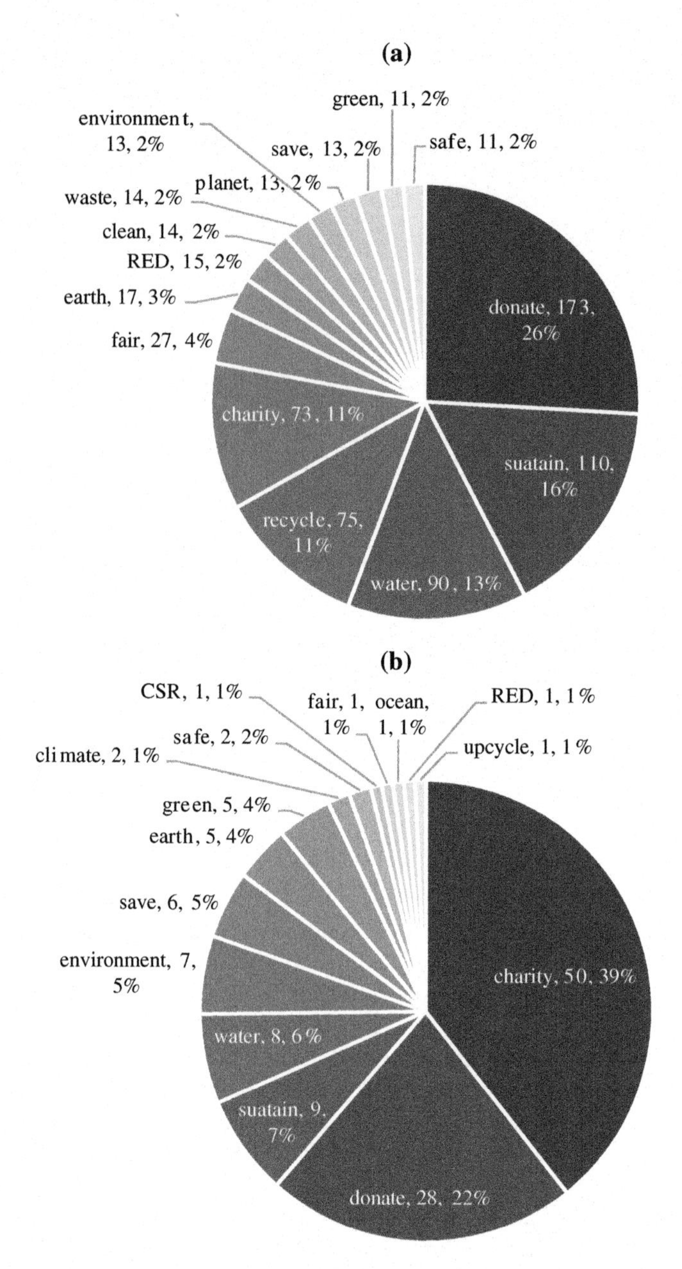

Fig. 8 **a** Distribution of top 15 mostly tweeted words (mass fashion brands). **b** Distribution of top 15 mostly tweeted words (luxury brands)

We further studied the usage and performance of multi-keyword combinations by mass fashion brands (46 combinations) and luxury fashion brands (16 combinations). Figure 12a presents the top 10 most tweeted word combinations for mass

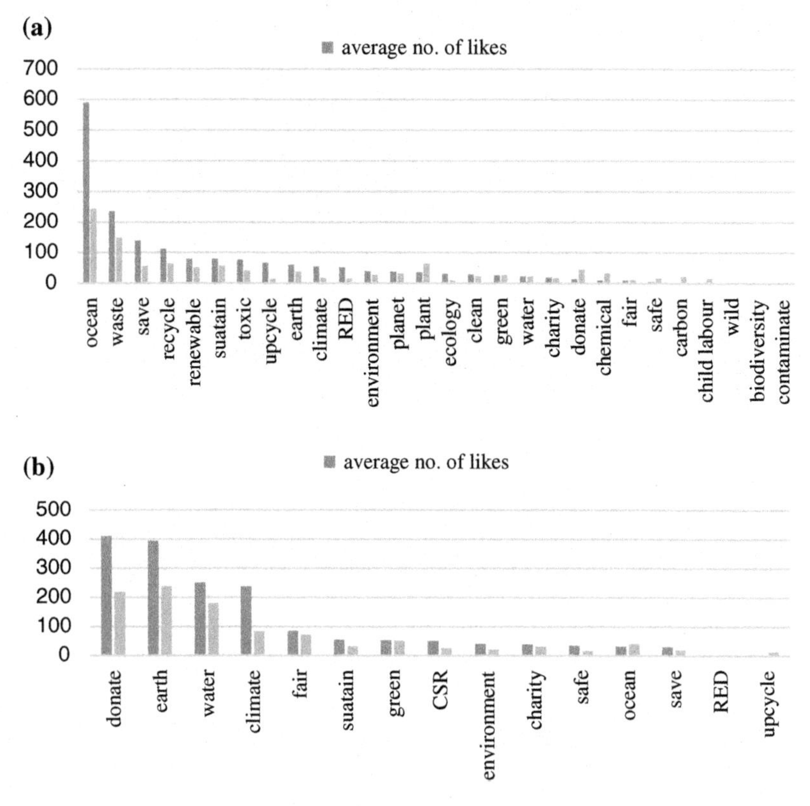

Fig. 9 **a** Average number of likes and retweets for the 28 keywords (mass fashion brands). **b** Average number of likes and retweets for the 15 keywords (luxury brands)

fashion brands. We found that mass fashion brands liked to tweet "recycle" and "plant," "clean" and "water," "donate" and "charity," "recycle" and "waste," and "water" and "donate" together. Figure 12b shows all 16 word combinations for luxury brands. All the word combinations were used only one or two times. It seems that luxury brands usually focus on one sustainability issue per tweet.

After studying the numbers of likes and retweets for mass fashion brands (Fig. 13a), we found that the results were similar to those of all brands, that is, the most liked multi-keyword combinations were not from the most tweeted combinations. For luxury brands (Fig. 13b), although they seldom used multi-keyword combinations, some still performed well (i.e., receiving a large number of likes and retweets), especially when "earth" and "water," "climate" and "earth," "save" and "donate," and "green" and "water" were tweeted together.

To better understand why some sustainability messages received more attention (i.e., large numbers of likes and retweets), we investigated in more detail the tweets with larger numbers of likes and retweets and found that the highest performing tweets contained # (hashtags) and @ (mentions). A hashtag is used to index topics

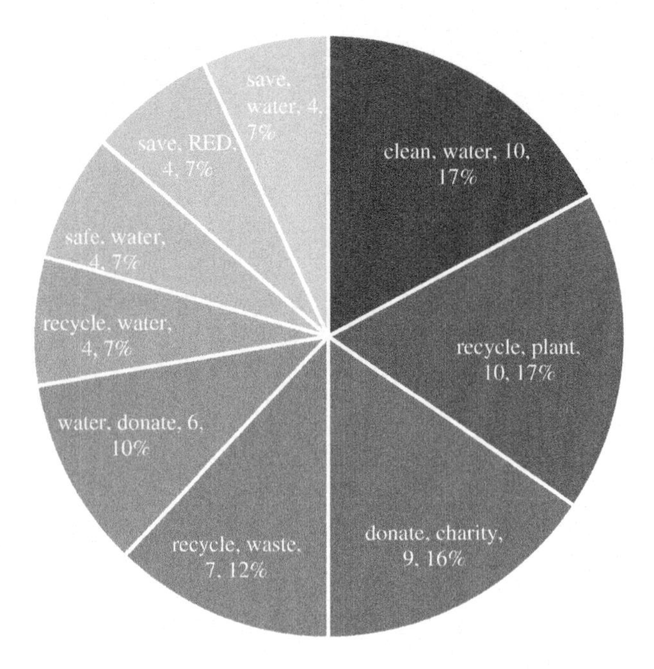

Fig. 10 Distribution of top 10 most tweeted word combinations (all brands)

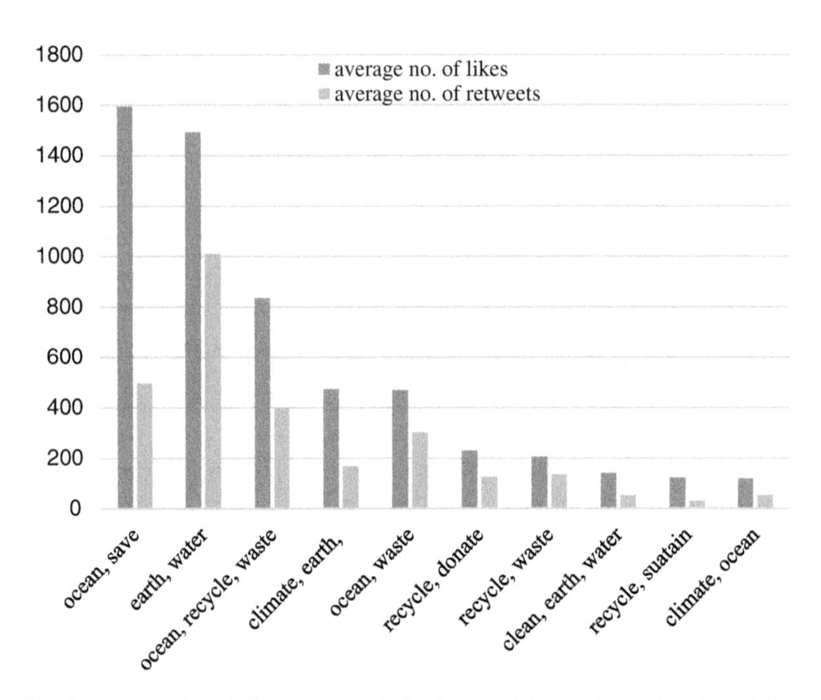

Fig. 11 Average number of likes and retweets for the top 10 keywords combinations (all brands)

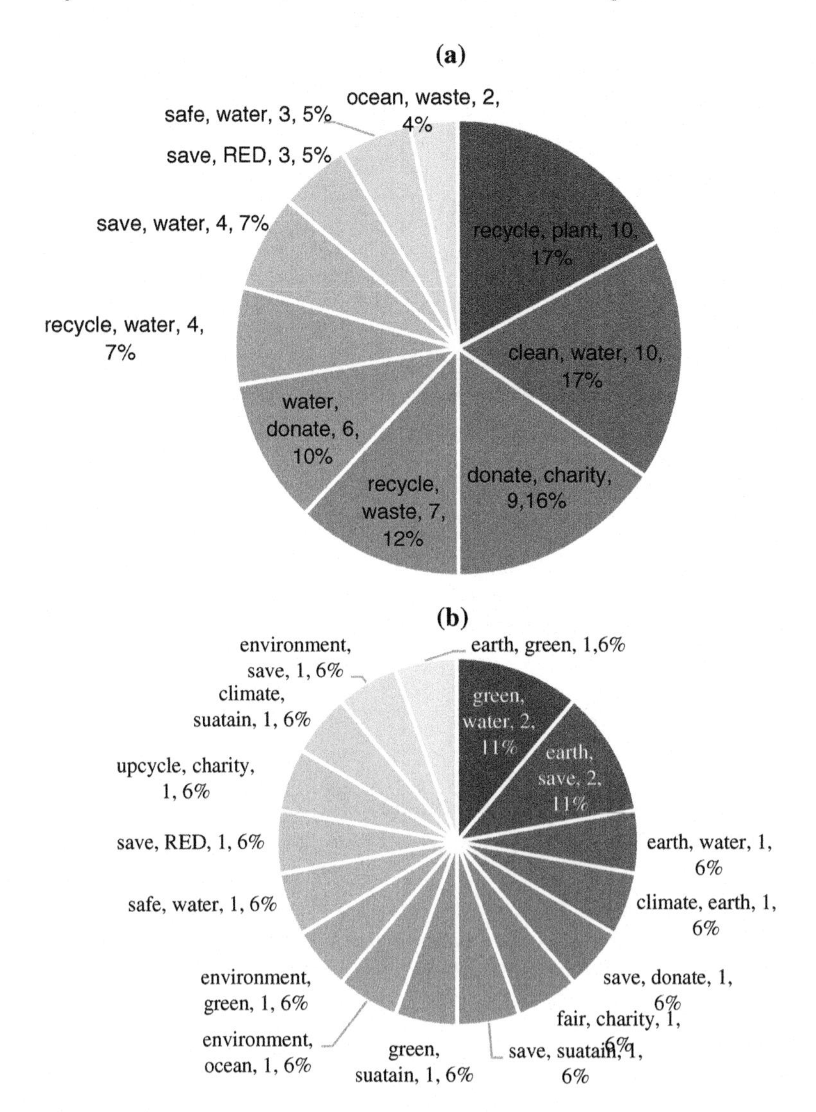

Fig. 12 **a** Distribution of top 10 most tweeted word combinations (mass fashion brands). **b** Distribution of all word combinations (luxury brands)

on Twitter, and a mention is a tweet that includes another user's account name in the tweet.[2] Hashtags and mentions are ways to link a Tweet to a topic (event) or a person. Brands' sustainability messages can be delivered effectively when linking to a popular topic (event) or a famous social figure. For instance, there were some

[2]For more detail of hashtag and mention, please refer to https://support.twitter.com/articles/49309 and https://support.twitter.com/articles/14023.

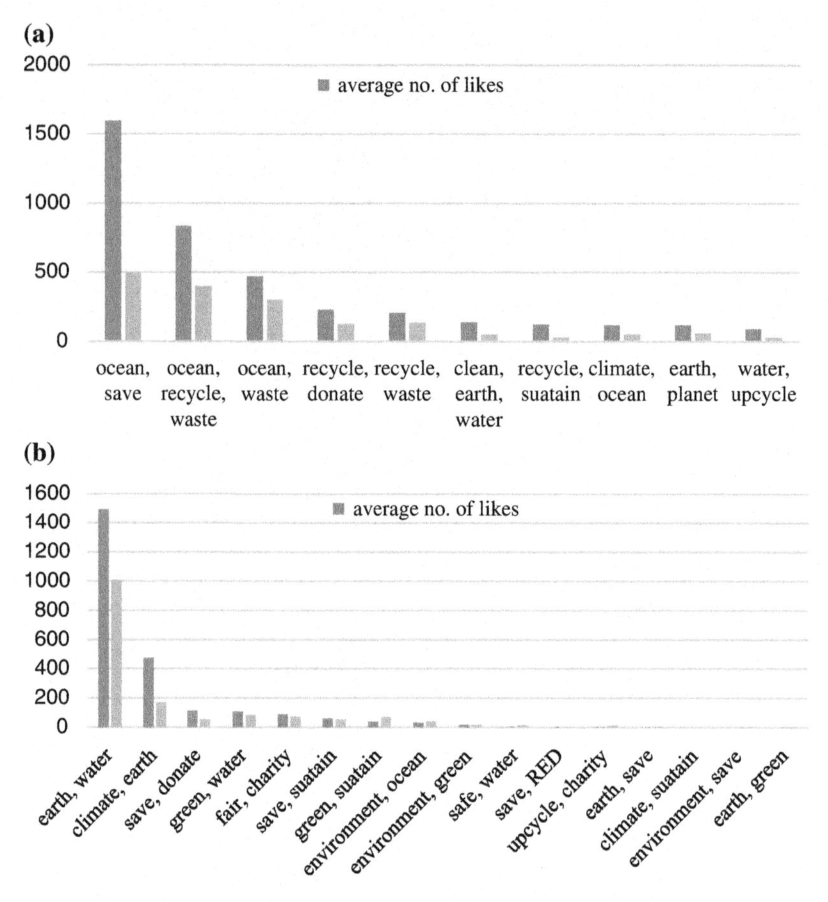

Fig. 13 **a** Average number of likes and retweets for the 10 keyword combinations (mass fashion brands). **b** Average number of likes and retweets for all keyword combinations (luxury brands)

brands' hashtags and mentions that were shown with an unusual number of likes and retweets, and those brands included Armani, Mark & Spencer, Coach, Adidas, and H&M. Armani's #EarthDay and #helpgivewater, M&S's #PlanA (a sustainable sofa produced by M&S), and H&M's #HMConsciousExclusive were highly liked and retweeted. Coach tweeted about its donation to an animal society initiative and mentioned @ArianaGrande, which received 10,878 likes and 5,430 retweets. All Adidas' tweets that mentioned @Parleyxxx received large numbers of likes and retweets. These tweets were relatively more popular than other tweets related to sustainability because of specific hashtags included in the tweets, such as #EarthDay and #HMConsciousExculsive. On the other hand, in the case of Coach's @ArianaGrande and Adidas' @Parleyxxx, the brands' tweets were related to a famous social figure or artist. As these social figures already had large numbers of followers, their sustainability tweets were highly recognized.

Conclusions

Given that there might be differences in consumer responses toward corporate sustainability messages between luxury and mass fashion brands because of distinctively different characteristics in their target consumers, this study explored consumer social media behavior and looked for potential differences in responses between the two groups of consumers. After analyzing over 89,000 tweets and retweets made by 11 luxury and 11 mass fashion brands as of March 2017, the study found several interesting results.

First, mass fashion brands were more active on Twitter than luxury fashion brands in general, yet luxury brands' messages were more often liked or retweeted by consumers. With regard to sustainability messages, the percentage of tweets, likes, and retweets was higher for mass fashion brands than luxury brands, suggesting consumers' strong support for mass fashion brands' sustainability efforts. However, when luxury brands do tweet about their sustainability efforts, they get three times more likes, or double the retweets of mass brands.

Second, the top 4 most tweeted words (usage frequency) by both mass and luxury fashion brands were "donate," "charity," "sustain," and "water." Specifically, mass fashion brands used more sustainability-related keywords, such as "sustain" and "water," and tweeted multiple keywords together, such as "recycle and plant," "clean and water," "donate and charity," and "recycle and waste." On the other hand, luxury fashion brands seldom used words, such as "recycle," "fair," "RED,", "clean," "waste," and "planet" together, and they usually focused on one sustainability issue in a single keyword tweet.

Overall, the analysis of Twitter messages suggests that luxury fashion brands are less communicative with consumers about their sustainability activities than mass fashion brands. Luxury fashion brands less often use the words "recycle," "fair," "clean," or "waste" as compared to mass fashion brands. Perhaps this is a response to luxury consumers' expectations for tackling sustainability aspects of luxury goods in a more suggestive and persuasive responses (Bendell & Kleanthous, 2007). Therefore, fewer tweet and less loud, yet effective in what they communicate might be luxury brands' strategies. Indeed, the findings do show that consumers look for leadership in luxury brands, as they tend to like or retweet more messages generated from luxury brands when their messages are focused on sustainability.

Yet, because of the sheer lack of such messages from luxury fashion brands, the Twittersphere seems dominated by mass fashion brands' Twitter messages. As much as negative messages or sustainability-related scandals are detrimental to luxury fashion brands, the findings also showed that positive messages improving overall sustainability are effective for generating consumers' positive word-of-mouth for luxury fashion brands. Therefore, the study's findings suggest that luxury fashion brands may want to be more intentional and strategic about their sustainability-related messages on social media platforms, as the characteristics of luxury fashion consumers could change.

This study does recommend additional research agendas. Because of the exploratory nature of these findings, further research is necessary to assess differences in social media consumer responses between luxury and mass fashion consumers. Next, an understanding of consumer profiles who "like" and "retweet" sustainability-related messages in both mass and luxury fashion brands is needed. That would help identify the opinion leaders of each market segment and use them proactively for brand management. Finally, this study recommends additional research on how to improve message contents themselves to be more memorable, remembered, and shared by core consumers of both mass and luxury fashion brands to help improve all fashion brands' communication efforts.

Appendix

Applied science	Cradle to cradle	Forced labor	Outreach	Stewardship
Awareness raising	CSR	Future	Partnership	Suburbanization
Balance	Culture	Gender	Photovoltaic	Supply chain
Behavior	Deforestation	Global	Planet	Sustainability
Biodiversity	Design	Globalization	Planning	Systems dynamics
Bioenergy	Development	Governance	Plant	Systems thinking
Biomass	Disaster	Green	Planting trees	Technology
Biomimicry	Diversity	Greenhouse gas	Policy	Three-legged stool
Brundtland Commission	Donate	Growth	Political	Three pillars
Building	Donation	Human condition	Poverty	Toxic
Carbon	Earth	Human rights	Preserving nature	Trade-offs
Carbon offset	Ecology	Human–environment interactions	Problem based	Transformation
Catalyze	Economic	Improving soil quality	Promoting biodiversity	Transit
Change	Efficiency	Innovation	Prosperity	Transparency
Change management	Emission	Interconnections	Public	Transparency enhancement
Charity	Energy	Interdisciplinary	Public health	Triple bottom line

(continued)

(continued)

Chemical	Engaged scholarship	Invest	Race	Underserved
Child care	Engagement	Justice	Recycle	Unintended consequences
Child labor	Entrepreneurship	Land	RED	Upcycle
City	Environment	Landscape	Reducing chemicals	Urban
Civil	Environmental support	Leadership in Energy and Environmental Design (LEED)	Reducing dyes	Urbanization
Clean	Equality	Life	Reducing packaging	Waste
Climate	Equity	Life cycle	Renewable	Water
Coastal	Externality	Limiting chemical use	Resilience	Water consumption
Collaboration	Fair	Local	Resources	Welfare
Commissioning	Fair compensation	Marine	Revolving fund	Wild
Community	Fair employment	Minority	Rights	Wind energy
Community health	Fairtrade	Modernization	Rural	Women
Community support	Farming	Movements	Safe	Working conditions support
Complex systems	Finance	Multidisciplinary	Save	
Conservation	Food	Natural systems	Sea level	
Conservation biology	Food chain	Nature	Social	
Conserve	Food systems	Nutrition	Solar	
Contaminate	Footprint	Ocean	Solutions	

References

Allwood, J. M., Laursen, S. E., Malvido de Rodríguez, C., & Bocken, N. M. (2006). *Well dressed? The present and future sustainability of clothing and textiles in the United Kingdom.* Cambridge: University of Cambridge Institute for Manufacturing.

Atasu, A. (2016). *Environmentally responsible supply chains.* Atlanta, GA, USA: Springer. https://doi.org/10.1007/978-3-30094-8.

Belk, R. (2014). You are what you can access: Sharing and collaborative consumption online. *Journal of Business Research, 67*(8), 1595–1600.

Bendell, J., & Kleanthous, A. (2007). *Deeper luxury: Quality and style when the world matters.* Retrieved November 30, 2016 from http://www.wwf.org.uk/deeperluxury.

Bennett, S. (2013). *Twitter was the fastest-growing social network in 2012, says study [STATS]*. Retrieved April 20, 2017 from http://www.adweek.com/digital/social-networks-growth-2012/.
Bhaduri, G., & Ha-Brookshire, J. (2011). Do transparent business practices pay? Exploration of transparency and consumer purchase intention. *Clothing and Textiles Research Journal, 27*(2), 135–149.
Bonner, S. (1997, February). Patagonia: A green endeavor. *Apparel Industry Magazine, 58*(2), 46–48.
Cachon, G. P., & Swinney, R. (2011). The value of fast fashion: Quick response, enhanced design, and strategic consumer behavior. *Management Science, 57*(4), 778–795.
Chae, B. (2015). Insights from hashtag #supplychain and Twitter analytics: Considering Twitter and Twitter data for supply chain practice and research. *International Journal of Production Economics, 165,* 247–259.
Correa, T., Hinsley, A. W., & De Zúñiga, H. G. (2010). Who interacts on the web? The inter-section of users' personality and social media use. *Computers in Human Behavior, 26*(2), 247–253.
da Silva Zago, G., & Bastos, M. T. (2013). Visibility of news items on Twitter and Facebook: Comparative analysis of the most replicated news in Europe and the Americas. *Brazilian Journalism Research, 9*(1 E), 114–131.
Dao, V., Langella, I., & Carbo, J. (2011). From green to sustainability: Information technology and an integrated sustainability framework. *Journal of Strategic Information Systems, 20*(1), 63–79.
Davies, I. A., Lee, Z., & Ahonkhai, I. (2012). Do consumers care about ethical luxury? *Journal of Business Ethics, 106*(1), 37–51.
De Vries, L., Gensler, S., & Leeflang, P. S. H. (2012). Popularity of brand posts on brand fan pages: An investigation of the effects of social media marketing. *Journal of Interactive Marketing, 26,* 83–91.
DEFRA. (2011). *Sustainable clothing roadmap: Progress report*. Retrieved April 20, 2017 form https://www.gov.uk/government/uploads/system/uploads/attachment_data/file/69299/pb13461-clothing-actionplan-110518.pdf.
Dhaoui, C. (2014). An empirical study of luxury brand marketing effectiveness and its impact a consumer engagement on Facebook. *Journal of Global Fashion Marketing, 5*(3), 209–223.
Dickey, M. (2014). Twitter gears up to launch a TweetDeck on steroids for Journalists. *Business Insider*. Retrieved on April 20, 2017 from http://www.businessinsider.com/twitter-and-dataminr-2014–1.
Einwiller, S. A., & Steilen, S. (2015). Handling complaints on social network sites—An analysis of complaints and complain responses on Facebook and Twitter pages of large US companies. *Public Relations Review, 41,* 195–204.
Finnerty, K., Stanely, B., & Herther, K. (2013). One green score for one earth: Industry report. *Market LOHAS*, 1–21. Retrieved April 20, 2017 from http://onegreenscore.com/images/OGS_Industry_Report.pdf.
Fletcher, K. (2012). Slow fashion: An invitation for systems change. *Fashion Practice, 2*(2), 259–266.
Fulton, K., & Lee, S. E. (2013). Assessing sustainable initiatives of apparel retailers on the internet. *Journal of Fashion Marketing and Management, 17*(3), 353–366.
Gallaugher, J., & Ransbotham, S. (2010). Social media and customer dialog management at starbucks. *MIS Quarterly Executive, 9*(4), 197–212.
Godey, B., Manthiou, A., Pederzoli, D., Rokka, J., Aiello, G., Donvito, R., et al. (2016). Social media marketing efforts of luxury brands: Influence on brand equity and consumer behavior. *Journal of Business Research, 69,* 5833–5841.
Goworek, H., Fisher, T., Cooper, T., Woodward, S., & Hiller, A. (2012). The sustainable clothing market: An evaluation of potential strategies for UK retailers. *International Journal of Retail & Distribution Management, 40*(2), 935–955.
Heil, O., Lehmann, D., & Stremersch, S. (2010). Marketing competition in the 21st century. *International Journal of Research in Marketing, 27*(2), 161–163.

Heine, K., & Berghaus, B. (2014). Luxury goes digital: How to tackle the digital luxury brand–consumer touchpoints. *Journal of Global Fashion Marketing, 5*(3), 223–234.
Hennings, N., Widmann, K.-P., Klarmann, C., & Behrens, S. (2013). Sustainability as part of the luxury essence: Delivering value through social and environmental excellence. *Journal of Corporate Citizenship, 52*, 25–35.
Henriques, A., & Richardson, J. (2013). *The triple bottom line: Does it all add up? Assessing the sustainability of business and CSR*. London: Earthscan.
Hulm, P., & Domeisen, N. (2008). Organic cotton. *International Trade Forum, 1*(2), 30–31.
Jansen, B. J., Zhang, M., Sobel, K., & Chowdury, A. (2009). Twitter power: Tweets as electronic word of mouth. *Journal of the American Society for Information Science and Technology, 60*(11), 2169–2188.
Janssen, C., Vanhamme, J., Lindgreen, A., & Lefebvre, C. (2014). The catch-22 of responsible luxury: Effects of luxury product characteristics on consumers' perception of fit with corporate social responsibility. *Journal of Business Ethics, 119*, 45–57. https://doi.org/10.1007/s10551-013-1621-6.
Joy, A., Sherry, J. F., Venkatesh, A., Wang, J., & Chan, R. (2012). Fast fashion, sustainability, and the ethical appeal of luxury brands. *Fashion Theory, 16*(3), 273–295. https://doi.org/10.2752/175174112X13340749707123.
Jung, S., & Ha-Brookshire, J. (2017). Perfect or imperfect duties? Developing a moral responsibility framework for corporate sustainability from the consumer perspective. *Corporate Social Responsibility and Environmental Management*. https://doi.org/10.1002/csr.1414.
Kim, J., Ferrin, D. L., & Rao, H. R. (2008). A trust-based consumer decision-making model in electronic commerce: The role of trust, perceived risk, and their antecedents. *Decision Support Systems, 44*(2), 544–564.
Kleanthous, A. (2011). Simply the best is no longer simple. *The Raconteur—Sustainable Luxury*. Retrieved April 20, 2017 from http://np.netpublicator.com/netpublication/n10444899.
Kozlowski, A., Searcy, C., & Bardecki, M. (2014). Corporate sustainability reporting in the apparel industry: An analysis of indicators disclosed. *International Journal of Productivity and Performance Management, 64*(3), 377–397.
Li, Y., Zhao, X., Shi, D., & Li, X. (2014). Governance of sustainable supply chains in the fast fashion industry. *European Management Journal, 32*, 823–836. https://doi.org/10.1016/j.emj.2014.03.001.
Malhotra, A., Malhotra, C. K., & See, A. (2012). How to get your messages retweeted. *MIT Sloan Management Review, 53*, 61–66. Retrieved April 20, 2017 from http://sloanreview.mit.edu/article/how-to-get-your-messages-retweeted/.
Marwick, A. E., & Boyd, D. (2011). I tweet honestly, I tweet passionately: Twitter users, context collapse, and the imaged audience. *New Media & Society, 13*(1), 114–133.
Park, H., & Lennon, S. J. (2006). The organizational factors influencing socially responsible apparel buying/sourcing. *Clothing and Textiles Research Journal, 24*(3), 229–247.
Riquelme, F., & González-Cantergiani, P. (2016). Measuring user influence on Twitter: A survey. *Information Processing and Management, 52*(5), 949–975.
Sadat, M. N., Ahmed, S., & Mohiuddin, M. T. (2014). Mining the social web to analyze the impact of social media on socialization. *Informatics, Electronics & Vision (ICIEV)*. https://doi.org/10.1109/iciev.2014.7135995.
Tsai, W. H. S., & Men, L. R. (2013). Motivations and antecedents of consumer engagement with brand pages on social networking sites. *Journal of Interactive Advertising, 13*(2), 76–87.
Wyatt, N. (2013). Best in class crisis management with social media. *Business 2 Community*. Retrieved April 20, 2017 from http://www.business2community.com/crisis-management/best-class-crisis-management-social-media-0658047.
Yan, R. N., Hyllegard, K. H., & Blaesi, L. F. (2012). Marketing eco-fashion: The influence of brand name and message explicitness. *Journal of Marketing Communications, 18*(2), 151–168. https://doi.org/10.1080/13527266.2010.490420.

Chapter 5
Thematic Analysis of YouTube Comments on Disclosure of Animal Cruelty in a Luxury Fashion Supply Chain

Heejin Lim

Abstract While consumers' increasing demands for luxury leather goods drives luxury brands to secure supplies of exotic and high-quality animal skins, animal welfare activists have attempted to raise the public's awareness of cruelty in the luxury fashion supply chains. Adopting the attribution theory, this study aims to uncover the pattern of consumers' responses to the issue of animal cruelty in the luxury fashion industry. Data were collected from consumers' responses to animal cruelty as revealed in PETA (People of Ethical Treatment of Animal)'s video on YouTube. Data analysis revealed distinct blame attributions and coping strategies, which depend heavily on viewers' attitude toward the video content. Findings from this study suggest that consumers' blame attributions are dispersed among different stakeholders, with luxury fashion brands and their customers treated as the causes of animal cruelty, slaughterhouse workers, and humans in general treated as perpetrators, and PETA and commenters that support PETA's message treated as accusers. Implications for the luxury fashion business and animal welfare promotion are discussed.

Keywords Animal cruelty · Luxury fashion supply chains · Luxury sustainability YouTube · Attribution theory · Thematic analysis

Global sales of luxury personal goods have been on the rise for the past decade (D'Arpizio, Levato, Zito, & de Montgolfier, 2015), and a strong and steady demand for leather goods account for a significant portion of this growth (D'Arpizio, 2016). For this reason, luxury brands compete to secure supplies of exotic and high-quality animal skins (e.g., reptiles and young ostriches) due to their aesthetic and rarity value (Wendlandt, 2013). Meanwhile, animal welfare activists have attempted to raise public awareness of cruelty in the supply chain of exotic and rare animal skins

---

H. Lim (✉)
The University of Tennessee, 1215 W. Cumberland Are, Knoxville, TN 37996, USA
e-mail: hlim@utk.edu

© Springer Nature Singapore Pte Ltd. 2018

C. K. Y. Lo and J. Ha-Brookshire (eds.), *Sustainability in Luxury Fashion Business*, Springer Series in Fashion Business, https://doi.org/10.1007/978-981-10-8878-0_5

for luxury brands. For example, in the USA, PETA (People for the Ethical Treatment of Animals) reported a case of animal cruelty in a crocodile farm in Texas (PETA n.d.). Their report revealed that crocodiles on the farm are raised in suffocating conditions for several months to years before being slaughtered for their skins. In this process, workers handle the crocodiles brutally and slaughter while the animals are still conscious and trying to escape. These slaughtered skins are sent to manufacturers of luxury brands and used to add allure to the brand's prestige.

Recently, conscious consumption has emerged as a mainstream social movement (Low & Davenport, 2007), and consumer goods companies have responded to this emerging consumer culture by competitively publicizing their ethical and environmental commitment. For example, H&M, a fast-fashion brand, announced that the company will use only recycled or sustainable materials in their production by 2030 and will increase sustainable operations throughout their entire value chain by 2040 (Dederich, 2017). In addition, after being accused of abusing human labor in a supply chain, Uniqlo, a Japanese fast-fashion retailer, announced their intention to enhance monitoring systems for the condition of their supplier factory and the well-being of their employees (McKevitt, 2017).

Although luxury brands have been relatively disengaged from sustainability issues compared to mass-market brands (Davies, Lee, & Ahonkhai, 2012), they do not seem to be exonerated from the responsibility for sustainability anymore (Winston, 2017). Lately, LVMH, a major luxury goods company, developed a sustainability program called LIFE (LVMH Initiatives for the Environment) that aims to pay close attention to sustainability issues in their supply chain and production processes (Winston, 2017). However, a close look at their green initiatives reveals that most of these initiatives focus on environmental issues such as energy consumption and potential pollution, while ignoring one of the most imperative issues for which luxury brands have been criticized—namely animal welfare in their supply-chain operations.

Similarly, the extant studies of sustainability in the fashion industry focus heavily on human well-being issues such as sweatshops (e.g., Phau, Teah, & Chuah, 2015) and environmental issues (e.g., Caniato, Caridi, Crippa, & Moretto, 2012), while animal welfare remains a relatively under-investigated topic (Molderez & De Landtsheer, 2015). Hennigs, Karampournioti, and Wiedmann (2016) argued that the practice of animal welfare is a critical part of corporate social responsibility, particularly in a company's sustainable sourcing practices. They proposed a conceptual model that depicts the conjoint effects of individuals' psychological traits (e.g., personality, empathy, self-identity, and moral concern) and situation-specific predictors (e.g., product involvement, ethical consumption value, and trade-off between ethical and conventional features in a product) on consumers' brand avoidance and boycott of unethical brands/products. While Hennigs et al.'s (2016) conceptual model propose individuals' internal factors as predictors of consumers' ethical consumption behavior, the current study focuses on consumers' causal attributional reasoning when they encounter a particular case of an unethical business process, namely animal cruelty.

This study adopts the attribution theory as a theoretical lens through which to view and interpret consumers' responses to the issue of animal cruelty in the luxury fashion industry. The theory posits that consumers investigate causes of an event and come up with several elucidations to explain that event (Folkes, 1984). This approach has been adopted widely in studying consumers' responses to a company's corporate social responsibility strategies (e.g., Yoon, Gürhan-Canli, & Schwarz, 2006), and attributions that consumers develop after a negative experience are found to influence their brand evaluations (Klein & Dawar, 2004). Davies et al. (2012) speculated that a prestigious and high-value image of luxury brands might mislead consumers to believe that those brands have a high level of morality and a low negative impact on society and the environment, which the authors call "the Fallacy of Clean Luxuries" (p. 41). Moreover, consumers might not be willing to trade-off the aesthetic and self-indulgent value of luxury goods against their moral intentions. Thus, the attribution approach is expected to help to explicate consumers' diverse views of ethics related to animal welfare in luxury fashion business. In brief, the purpose of this study is to enhance understanding of the following proposed research questions: (1) how do consumers respond when they are exposed to a sustainability issue, particularly animal cruelty in luxury brands' sourcing practices? (2) How do consumers attribute blame for animal cruelty in a luxury fashion supply chain?

Animal Welfare in the Luxury Fashion Business

> Unto Adam also and to his wife did the LORD God make coats of skins, and clothed them.
>
> (Genesis 3:21, King James Version)

Leather was the first garment worn by humankind, and it has retained its prestigious status even though later advancements led to the creation of diverse natural fabrics such as cotton and silk. Undeniably, the label of genuine leather has been recognized as an indicator of high quality and nobility throughout many different eras (Sterlacci, 1997). The popularity of leather in the luxury fashion business increased in the 1960s, when major high-street brands such as Hermès, Nina Ricci, and Yves Saint Laurent introduced beautifully crafted leather apparel into their collections. The further advancement of the technique enabled designers such as Giorgio Armani, Claude Montana, Emanuel Ungaro, and Vivienne Westwood to create innovative and marketable leather clothing in the 1980s. Owing to the rise of women's economic status, a global demand for leather surged in the luxury fashion industry in the same period, as the leather handbag gained importance and became a symbol of luxury. Accordingly, efforts to secure high-quality leather and novelty skins have become more and more competitive among luxury goods makers to create exclusive looks (Sterlacci, 1997). Sterlacci (1997) claimed that an animal's condition at the time of being killed affects the quality of leather palpably. For example, the greater an animal's pain and physical resistance are, the lower is the quality of leather.

Most leather for the luxury fashion industry is supplied by slaughterhouses. Originally, animal skins for clothing were produced as a by-product of food consumption. As the popularity of leather goods rose after the French Industrial Revolution in the nineteenth century; however, public slaughterhouses were developed with the sole purpose of supplying animal skins for human use (Brantz, 2005). It has been argued that the institutionalized system of slaughterhouses has generated great insensibility to the quality of animal life, such that animal cruelty has been committed covertly at a large scale (Otter, 2008). Meanwhile, consumers have been unwilling to confront ethical culpabilities in the process of slaughtering animals for consumption (Williams, 2008). Williams (2008) characterized this phenomenon as "affected ignorance," which reflects humans' purposive choice to ignore an important moral issue that involves significant pain to victims as a result of their actions (p. 372).

However, recent consumer culture signals consumers' increasing demand for transparency about animal products they consume (The Hartman Group, 2015), and companies have no choice but responding to this trend. For example, cosmetic companies that were previously well-known for their cruel treatment of animals in product testing have proactively introduced marketing campaigns to assure consumers of their ethical standards in production (Hennigs et al., 2016). In addition, companies such as Hilton and McDonald promote their sustainability initiatives in their global food supply chains. While the standards of animal welfare have been heavily promulgated in food supply chains (e.g., Brantz, 2005) and the cosmetic industry (Hennigs et al., 2016), there is a dearth of studies investigating animal welfare in supply chains of the luxury fashion business.

Attribution Theory and Corporate Social Responsibility

Attribution refers to "the perception or inference of cause" (Kelley & Michela, 1980, p. 458). Attribution theory examines the cognitive processing of an information recipient that generates causal attributional reasoning when provoked by others' behavior in social environments (Nyilasy, Gangadharbatla, & Paladino, 2014). Kelley and Michela (1980) explained that one's response to others' behavior is determined by how the receiver interprets the cause of the behavior. According to Laczniak, DeCarlo, and Ramaswami (2001), causal attribution is formed based on three components of the information: the stimulus, the communicator, and the circumstance.

As a response to the given information (Laczniak et al., 2001), attributional reasoning is found to be evident when consumers evaluate companies' sustainability messages and performance. For example, Ellen, Webb, and Mohr (2006) found that consumers process a company's claim of corporate social responsibility (CSR) by inferring the company's motives. Instead of simply accepting or rejecting the CSR claim, viewers engender attributions of the company's motives for engaging a CSR initiative. Ellen et al.'s findings (2006) revealed that consumers are likely to respond to the claim positively when they identify the company's motive to be strategic and values-driven, while they will respond negatively to egoistic and stakeholder-driven values. In addition, researchers suggested that consumers' attributional reasoning is more evident in negative consumption experiences (Hess, Ganesan, & Klein, 2007; Weiner, 2000). Thus, the attributional process mediates the effects of marketing information on consumers' responses to a company's message (Ellen et al., 2006), attitude, and brand evaluation (Weiner, 2000), and behavioral intention (Yoon, 2013).

Methods

Data were collected from consumers' responses to animal cruelty as revealed on YouTube, a video-sharing social networking site. YouTube, which has attracted more than a billion users worldwide to date (YouTube n.d.), allows users to post comments on content and interact with one another. Researchers have used YouTube to study various topics such as the effectiveness of consumer-generated advertising (Lawrence, Fournier, & Brunel, 2013) and a content analysis of fat stigmatization in YouTube videos (Hussin, Frazier, & Thompson, 2011).

For this study, the author investigated videos of animal cruelty in the supply chain of luxury fashion brands and selected a PETA's YouTube video that shows the process of the killing of baby ostriches in a slaughterhouse which supplies the baby ostrich skins to luxury fashion brands such as Prada and Hermès. The video contains horrific images and stories of young ostriches being butchered inhumanely:

> Narrator: Workers forced him [a three-month old ostrich] into a box to be electrically stunned; many of the panicked birds slip and fall on the floor while being manhandled into the machine… workers clamp their legs and bodies to immobilize them and stick their heads into the stunner…

This content was selected because of the richness of data (i.e., the quantity of viewers' comments) that it offered. Posted on February 2016, the video has received more than one million views and 322 comments, and the frequency of new comments had decreased by the time of data collection. To increase reliability, comments on the same video posted by PETA UK were also collected and analyzed. The PETA UK video was posted on the same date and had attracted almost three million views and 557 comments by the time of data collection. In total, 879 comments comprising 20,691 words were copied and pasted into a Word file and imported to NViVo 11 for analysis. Comparison of the two data sources verified no significant difference between the data from the PETA US and PETA UK accounts in terms of code generations. Additionally, consumers' responses to the same video posted on Facebook were used to verify identified codes from the YouTube data. Out of 7,058 postings at the time of data collection, PETA's replies to consumers' postings were excluded from the analysis. The identified codes from the YouTube data were sufficiently representative of consumers' responses on Facebook.

The Facebook data were used to add richness to the quotes used to illustrate each identified theme.

This study adopted thematic analysis to analyze comments on the YouTube video. Thematic analysis is "a method for identifying, analyzing and reporting patterns (themes) within data" (Braun & Clarke, 2006, p. 79). Themes that describe viewers' responses to animal cruelty in the videos were identified by the iterative process of thorough reading and re-reading of the data (Rice & Ezzy, 1999). This study employed a hybrid approach of thematic analysis that involves both the inductive approach (in which the identification of themes derives heavily from the data; Boyatzis, 1998) and the thematic approach (in which coding is driven by a researcher's theoretical framework—i.e., blame attributions; Braun & Clarke, 2006).

Thematic analysis was conducted in multiple stages, as suggested by Braun and Clarke (2006): (1) the researchers read and re-read each comment to familiarize themselves with the data; (2) researchers generated initial codes systematically across the entire data set; (3) themes were extracted by collating codes into potential themes (i.e., first-order themes; Fereday & Muir-Cochrane, 2006); (4) themes were refined based on internal homogeneity and external heterogeneity (Patton, 1990); and (5) final themes were defined and labeled. The iterative process of comparing and linking codes and themes then proceeded to interpreting and generating an illustrative framework that is representative of the responses. A second judge was recruited to ensure the validity of data interpretation and conducted thematic coding independently based on the author's coding schemes. Two judges discussed coding agreements, disagreements, and further modification.

Findings

This section provides the results of the thematic analysis and interprets findings. Two overarching themes were identified in viewers' responses to the animal cruelty videos on YouTube and Facebook: (1) blame attributions and (2) coping strategies. Findings revealed that properties of these two themes are strongly associated with whether viewers' attitudes toward the video were receptive or antagonistic. The receptive viewers indicate those who agree with PETA's claim of animal cruelty in the video, while antagonistic viewers indicate those who do not agree with the communicator (i.e., PETA)'s intention in the video content. Depending on viewers' attitudes, the comments presented distinct targets of blame and patterns of how to cope with the revelation. This finding supports Kelley and Michela's notion (1980) that a receiver's interpretation of the cause determines his/her response to others' behavior. In this study, the first theme of blame attributions pertains to viewers' perception of the cause of animal cruelty, and the second theme of coping strategies illustrates the viewers' responses to the reported animal cruelty as a result of such attributional reasoning. Figure 5.1 presents a general framework of YouTube viewers' responses to animal cruelty revealed in a luxury fashion supply chain.

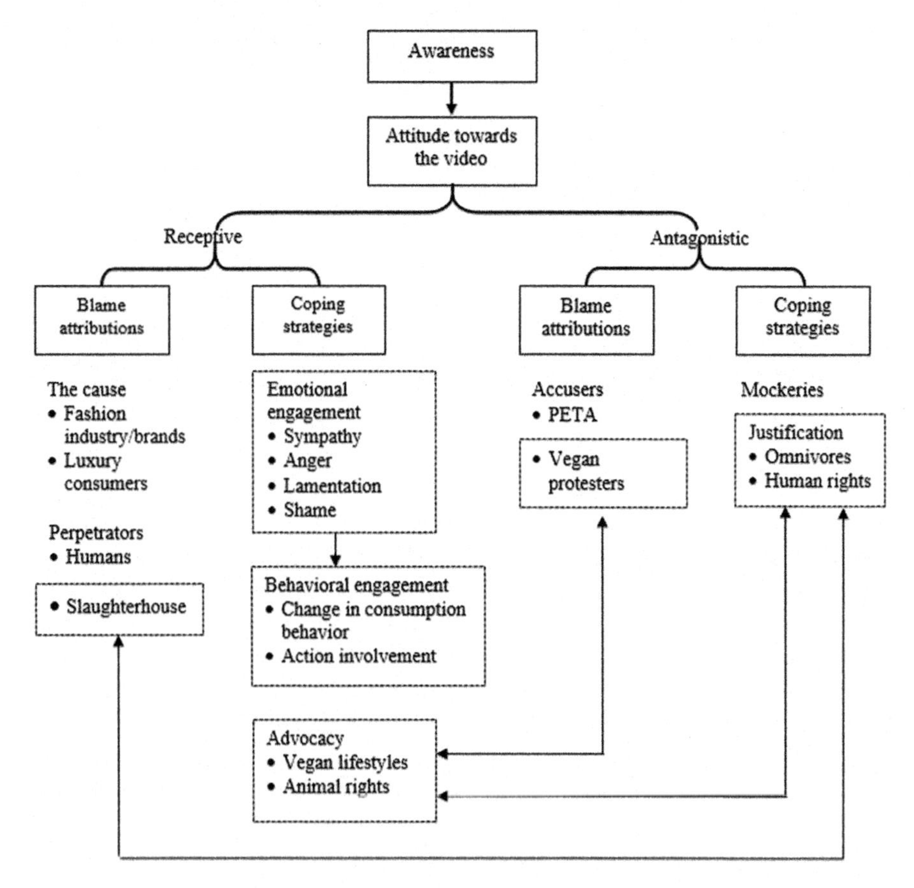

Fig. 5.1 General framework of consumers' responses to a YouTube video depicting animal cruelty in a luxury fashion supply chain. *Note* One-way arrows indicate the pattern of a viewer's information process; two-way arrows indicate the progress pattern of two-way discourses among viewers

Blame Attributions: Receptive Viewers

Two major targets of blame were identified in comments from receptive viewers: (1) the cause and (2) the perpetrators of animal cruelty in the slaughterhouse. First, viewers blamed the fashion industry and luxury brands (i.e., Prada, Hermès) and people who buy their products for being the cause of animal cruelty in the slaughterhouse. Although these groups are not directly involved in the killing process directly, the viewers perceived that they contribute to the cruelty indirectly by sourcing and seeking the animal skin. Statements such as "killing beautiful animals for a stupid bag and clothes" and "if you pay a fortune for a bag that caused the inhuman torture of these young birds …" indicate that viewers identify associated brands and buyers of their products as a driving force of the brutality. Below are examples of quotes for this category of blame attributions.

> "Tragic murder of these wonderful birds just for "luxury goods". It is disgusting."

> "Those bags are hideous, symbols of animal cruelty. Thanks for this video exposing the atrocious cruelty behind these 'luxury' products."

> "Hermess and Praada you are so evil."

> "wtf… im sooo disgusted. Looking at fucking ratchet girls carrying a hermes or a prada bag around n proud that they own something so expensive makes me sick to my stomach. Even though if it weren't vegan, spending 3000 dollars on a fucking handbag is bullshit, the money for 1 bag could probably do so much 2 help the world."

> "these days woman's fashion looks terrible but they just wanna get noticed cause their parents & friends never notice them i understand they have inferior complexity, but buying these products is waste of life and money."

The second target of blame from receptive viewers involves the perpetrators who commit the mistreatment of animals. The first group of perpetrators includes workers in the slaughterhouse. Viewers blamed the workers by calling them "damn murderers" and "evil bastards." In addition, the data revealed that viewers blame not only workers but also the inhumane system in the slaughterhouse. In these comments, people blamed the slaughterhouse's ignorance of animals' suffering and pain in the killing process. The following quotes illustrate viewers' blame attribution to the slaughterhouse workers and the slaughtering process.

> "They [workers] are the scum of the earth. Who gives them the right to take the life of an innocent Angel. Bastards should not be walking on this earth. SICK VILE EVIL BASTARDS. THERE IS A SPECIAL PLACE IN HELL FOR THIS SCUM."

> "How could the workers be so soulless?"

> "it's not only eating animals it's abusing them. Do you like being hit? No. how about 10 times harder. THIS IS WHAT ANIMALS LIKE THIS GO THROUGH. That's what's wrong with it."

> "The point is we try to limit the amount of suffering we cause… to all life."

The data also suggested that some viewers stretch their definition of perpetrators by blaming human beings in general as perpetrators. Although workers of the slaughterhouse are presented as direct abusers of the animals in the video, a substantial number of comments expressed lamentation for cruelty in human nature. As presented in the following quotes, these viewers identify human beings in general as perpetrators of animal cruelty who lack respect for animals and nature.

> "I'm ashamed to belong to Humanity. People are the most horrible creatures on the world… There is no God, there exists only one failure in Nature and it's called humans."

> "We humans are disgusting animals"

Coping Strategies: Receptive Viewers

Data analysis revealed diverse patterns of viewers' coping strategies after being exposed to animal cruelty in a luxury fashion supply chain. Coping is defined as "cognitive and behavioral efforts to manage (master, reduce, or tolerate) a troubled person-environment relationship" (Folkman & Lazarus, 1985, p. 152). The YouTube and Facebook data showed that the first stage of receptive viewers' coping involves emotional engagement. Specifically, people who viewed the video of animal cruelty in the slaughterhouse sympathetically expressed negative emotions such as "disgusting," "horrifying," "shocking," "angry," and "sad."

Often, viewers' negative valence motivates them to engage in problem-focused coping that involves direct actions or changes in behavior to resolve the situation that causes their negative emotions (Folkman & Lazarus, 1985). This study identified various coping strategies that people used when they encountered evidence of animal cruelty in a luxury fashion supply chain. The first type of problem-focused coping involves taking action to stop the cruelty. For example:

> "I'll happily sign any petition if there is one. I hope they are making a difference when we do sign them!"

> "Signed [a petition]. Hermes/prada and all the other designer companies that do this need boycotting!"

> "Sadly, there is no way to stop this unless we actually do something, not just sit here and watch it, and make a sad face and say, I wish this would stop. I, for one, TRY to make an effort toward helping, not just saying I will. No offense to anyone once so ever."

In the same vein, problem-focused coping appeared in the way that some viewers sought to join the activity to stop animal cruelty. For example, "now how can [I] become a animals rights activist?" "can you link the petition to sign please?" "Can I volunteer for PETA UK? And how?"

In addition, viewers' behavioral engagement involves changes in consumption behavior at the individual level, such as avoiding a future purchase of a product from associated brands such as Prada and Hermès or even extreme behavioral changes to avoid any animal products. For example:

> "i will never purchase from all these luxury brands again, i have multiple pieces from hermes, prada etc. (not ostrich leather) but i will not aid in continuing this cycle."

> "I never knew that!! Never. Buying anything made out of animal."

> "This is so WRONG!!! I will never buy any products that harm animals. Sharing!"

Problem-focused coping was also evident in the form of advocating animal rights and vegan lifestyles. For example:

> "I see no problem with killing animals for absolute survival. However, humans living in 1st and most 2nd world countries do not need to eat animal products at all, and in terms of health, would benefit without them. There are numerous vegan products to replace animal products. Even without these alternatives, there is still no excuse to not be vegan if there is

> an abundance of fruit and vegetables (which in most countries there is)… Humans… do not need to consume animal products to be healthy. We can live long and healthy lives eating plants alone."

> "world hunger can be solved by veganism. 1/3 of all our plant food is fed to animals in farms. If we are plants only, there would be enough food for everyone."

Data analysis unveiled that such advocacy of the extreme option of going vegan was often generated in response to comments that blamed PETA and justified meat eating, leading to battles of blame between vegan advocates and meat eaters.

Blame Attributions: Antagonistic Viewers

In the YouTube and Facebook data, a considerable number of comments are hostile to PETA's video. The main targets of blame for these comments include the accusers of animal cruelty, such as PETA and vegan protesters. The data analysis revealed that this group of people criticized PETA for various reasons. For example, some people blamed PETA for its intention to reveal the animal cruelty. As evidenced in the following quote, this group of commenters disputes PETA's claim for animal cruelty in a slaughterhouse and argues the legitimacy of PETA's cruelty claim.

> "PETA should stop making this ridiculous video, i mean WTF?! those ostrich are bred for their specific purpose just like chickens and cows and goats and pigs. PETA is just sucker for the donations."

> "If butchering the ostriches for leather goods is bad, what about butchering the other animals for food? Chicken, cow or the lamb? This is just a media propaganda."

> "ummm…there's people slaughtering innocent men, women and children all over the globe. Billions of people are starving, and the world is on the brink of WWIII, and you guys spend all this time and money on farm raised ostriches????"

Antagonistic viewers' blame attribution to PETA is often expanded to the group of vegan protesters in general. As evidenced by the following quotes, this type of blame is strongly related to one's position regarding the human right to consume animal products.

> "These are generation of birds whose parents were raised in a farm that had babies that grew up and had more babies… These companies don't waste time going out to trap thousands of birds each day… So don't give me that shit that they are being taken away from their habitat… These birds don't know anything outside of a farm… My god its so stupid how you hippies keep comparing humans to a fucking dumb bird"

> "They are not endangered species. Slaughterhouse is looking pretty neat to me… I eat meat and I love it. Please do not force people to be like you. Diet is like religion. If you force it you will be sorry later."

> "… animals were created to benefit humans, you are free to keep a lamb as a pet but at the same time it is just not right to blame others who eats lamb meat."

In addition, these people criticized vegan advocates for being "hypocritical" and for "judging" others. These comments were often generated in response to claims of the superiority of a vegan lifestyle.

> "Ugh, I hate vegans. Every vegan I've ever met has been a self righteous, self important, stupid privileged millennial. I lived with one for a year when I was 18, and she'd lecture me CONSTANTLY about how I was a murderer, an animal nazi, how I was destroying the planet, and when we had cockroaches she wouldn't fucking kill them!!! What a mong."

> "why you judge people who eat meat? You may be don't eat meat but what about cheese or other dairy product or what you wear on your shoes the leather of your valet your belt and your leather jacket or your working safety boots?"

Coping Strategies: Antagonistic Viewers

While receptive viewers' coping involves different ways of taking action to cope with their disturbed emotions aroused by the animal cruelty video, antagonistic viewers' coping tends to be emotion-laden, likely due to their efforts to diminish their experienced emotions caused by their perception of being accused (Luce, 1998). By the video content and others' blame attributions, these people perceive their non-vegan lifestyle to be threatened, and they cope with this stressful situation in various ways. The first type of coping is mockery. These comments deride PETA and vegan advocates by showing a firm attitude in support of a non-vegan lifestyle.

> "This video made me want to try Ostrich meat."

> "Now i just want an Ostrich bag!"

> "ostrich burgers slap Yum i want one…"

In addition, the data analysis revealed that antagonistic viewers' emotion-focused coping involves strong justification of their non-vegan lifestyle and humans' right to control and consume animals. These people appear to interpret the video and others' receptive comments as blaming people who lead a non-vegan lifestyle and consume animal products such as food and clothing. Thus, they attempt to dispute PETA's animal cruelty claim and others' supportive comments by justifying their non-vegan lifestyle.

> "You are making the mistake of attributing human emotions and nervous system to the animal kingdom. ALL creatures are not 'wired up' or 'programmed' in the same way as humans. We are unique in our qualities, and our emotions and nerves are very highly evolved, way beyond animals."

> "…humanity has been eating meat for millions of years and 97% of the population eats meat. You cannot change it because that is our natural diet. Your religious vegan ideology will never change that."

> "Animals are a gift for us to use for food, clothing, and education. Don't be a fool for PETA."

Discussion

Based on the attribution theory, this study aims to uncover patterns of consumers' responses to animal cruelty in a luxury fashion supply chain, as revealed on popular social media platforms (i.e., YouTube and Facebook). Previous studies have identified diverse values that motivate consumers to purchase a luxury brand. For example, (Wiedmann, Hennigs, & Siebels, 2007) suggested that consumers' luxury perception is associated with individual values such as self-pleasure as well as interpersonal values such as impression management. In addition, Vigneron and Johnson (2004) identified conspicuousness, uniqueness, and quality as indicators of a non-personal dimension of luxury perception and hedonism and extended-self as an indicator of a personal dimension of luxury perception. As such, the extant literature focuses heavily on consumers' motivation to purchase luxury products based on the products' financial, functional, individual, and social value, and consumers' demand for sustainability in this sector has been underexplored.

The Importance of Awareness

The literature shows that an ethical process of production is relatively neglected in luxury consumption compared to commodity purchases (Davies et al., 2012). A possible explanation for this might be found in Achabou and Dekhili's (2013) claim about conflicting values between luxury and sustainability. That is, while luxury tends to be associated with its hedonic, materialistic, and conspicuous value, sustainability involves contrasting values such as altruism, abstinence, self-control, and morality. In addition, a previous study showed that consumers do not want to sacrifice quality and design at the cost of sustainable apparel products, despite their positive attitudes toward green issues (Crane & Clarke, 1994).

However, findings from this study suggest that consumers' negligence of ethical concerns in luxury consumption can be attributed to a lack of awareness. In the YouTube and Facebook data, a substantial number of responses showed that they had been previously unaware of animal cruelty in a luxury fashion supply chain (e.g., "but I had no idea they were slaughtered as well," "huge eye opener," "I never knew that!!") Often, this new awareness elicited negative emotions and led to the intention to engage in problem-focused coping, such as future avoidance of purchasing associated brands and/or animal products and participating in collective action to stop animal cruelty in the supply chain.

Blame Battles: Emotion-Laden Debates

Findings from this study illustrated severe blame battles between vegan advocates and meat eaters in response to the animal cruelty video. As presented in Fig. 5.1, claims of the superiority of a vegan lifestyle often raise animosity and ignite aggressive disputes from non-vegan people, and vice versa. Often, these debates become emotional and offensive, as evidenced by the use of offensive language in the following quotes:

> "I didn't know meat eating could lead to stupidity."
>
> [Responses to the comment above]
>
> "Animals kill other animals for food. It is not murder you ignorant tard."
>
> "yeah imagine your flayed skin ending up in someone's luxury bag."
>
> "You stupid bitch."

As these quotes suggest, when the crux of the issue becomes consumers' choice between a vegan and non-vegan lifestyle, non-vegan consumers undertake a protective coping approach. A previous study suggested that consumers tend to resist negative word-of-mouth for their preferred brands (East, Hammond, & Lomax, 2008). In this study, consumers who maintain a non-vegan lifestyle show resistance and animosity toward the video content and others' vegan claims when they interpret the message as a threat to their beliefs and lifestyle. As a result, they refuse to alter their attitude and change their behavior.

Implications for Luxury Fashion Business

Findings from this study suggest that consumers' blame attributions are dispersed among different stakeholders, with luxury fashion brands and their customers treated as the causes of animal cruelty, slaughterhouse workers, and humans in general treated as perpetrators, and PETA and commenters that support PETA's message treated as accusers. Although luxury fashion brands are not direct perpetrators of violence in a supply chain, findings indicate that consumers identify the associated companies as a responsible cause of animal cruelty.

Although luxury brands have been exempted from the use of sustainable practices of animal welfare compared to mass-market brands or other industries, findings from this study suggest that the luxury industry should nonetheless pay attention to consumers' increasing demands for ethical business practices and transparency. In particular, in the era of social media, the effects of word-of-mouth are immeasurable, and luxury brands cannot be exempted from the sustainability issue by their intrinsic quality and rarity value (Achabou & Dekhili, 2013). While previous studies have identified a gap between consumers' attitudes toward ethical consumption and their actual behavior (e.g., Carrigan & Attalla, 2001; Moraes, Carrigan, & Szmigin, 2012), findings from this study suggest that consumers develop attributional reasoning to a great extent when they become aware of an ethical breach in a luxury fashion supply chain, which leads to their behavioral engagement to resolve the situation.

Marie-Claire Daveu, chief sustainability officer for the Kering luxury group, highlighted the luxury industry's responsibility for sustainability as a trendsetter and innovation leader. Luxury brand companies should take this issue seriously not only to avoid consumers' criticism or potential marketing loss but also in a genuine effort to exercise a positive impact on the world. To do so, it is recommended that companies producing luxury consumer goods hold a high standard and take pride in their workmanship, which should also include a careful and ethical process of material sourcing.

Implications for Animal Welfare Promotion

Findings from this study demonstrated the effectiveness of social media as a means of raising awareness of animal welfare issues. The high level of openness and accessibility on social media has promoted a consumer culture in which brand transparency is key in consumer-brand relationships. Thus, animal welfare activists and organizations should utilize social media platforms as a strategic communication center that raises awareness and builds networking.

In turn, promoters of animal welfare should pay close attention to how to position their claim and make it clear in their message. As previously mentioned, this study demonstrated that consumers use defense mechanisms to resist sustainability issues when the message, from either a main communicator or supporters, is perceived to threaten their own beliefs and lifestyles and compel them to choose an extreme end. For example, a description of PETA's YouTube video shows that the communicator's intention is to stop animal consumption. This is also evident in their petition statement. Findings from this study suggest that such pressure generates immediate animosity among non-vegan consumers. In this case, they seal off their attention from ethical conditions of production and raise a dispute that attributes blame to the accusers of animal cruelty. In this case, the message fails to lead non-vegan consumers to interpret a sustainability message as intended and engage in behavioral actions.

In addition, findings from this study revealed that blame battles between receptive viewers and antagonistic viewers are frequently attributed to a discord in the crux of the issues. While receptive viewers attempt to focus on the ethics of the slaughterhouse process, antagonistic viewers tend to translate PETA's message as an accusation toward animal consumption. In this case, these dialogues tend to end in emotional weariness. Thus, it is recommended that animal welfare activists and organizations promulgate their campaign with a clear focus in their messages to educate consumers regarding the importance of ethical conditions of production and the need to develop rigorous standards in material sourcing.

Conclusion

> "Dominion does not mean destruction, but responsibility. It is important to avoid flawed convictions about the right and power of humankind in relation to the rest of the natural world."
>
> -Holcomb

Today's consumers' care about what they wear and what is in the products they use. That is, consumers choose a company not only for its products but also for the company's mission, values, and efforts to make the world a better place (Kotler, Kartajaya, & Setiawan, 2010). Animal welfare and the slaughtering process have been relatively under-investigated topics in research on sustainability issues. However, significant changes in cultural attitudes toward slaughtering and increasing sensibilities between humankind and animals call on luxury fashion businesses to pay attention to their sustainable supply chain in sourcing animal skins (Fitzgerald, 2010). Analyzing social media data of consumers' responses to animal cruelty in a luxury fashion supply chain, findings from this study make a unique contribution by shedding light on diverse targets of consumers' blame attributions and patterns of consumers' coping with an animal welfare issue in the luxury fashion industry.

It should be noted that, although data collected from popular social media platforms such as YouTube and Facebook provide valuable insights about patterns of consumers' responses to animal cruelty in a luxury fashion business chain, the data also present limitations. First, the social media data do not enable researchers to probe into further meanings and discourse. Second, because of the limited exposure of commenters' profiles, it is difficult to examine relationships between their personal profiles and their responses. Therefore, future study is encouraged to explore these factors by employing various research methods in order to extend the understanding of consumers' responses to a sustainability issue.

References

Achabou, M. A., & Dekhili, S. (2013). Luxury and sustainable development: Is there a match?. *Journal of Business Research, 66*(10), 1896–1903.

Boyatzis, R. E. (1998). *Transforming qualitative information: Thematic analysis and code development*. Thousand Oaks, CA: Sage Publications.

Brantz, D. (2005). Animal bodies, human health, and the reform of slaughterhouses in nineteenth-century Berlin. *Food and History, 3*(2), 193–215.

Braun, V., & Clarke, V. (2006). Using thematic analysis in psychology. *Qualitative research in psychology, 3*(2), 77–101.

Caniato, F., Caridi, M., Crippa, L., & Moretto, A. (2012). Environmental sustainability in fashion supply chains: An exploratory case based research. *International Journal of Production Economics, 135*(2), 659–670.

Carrigan, M., & Attalla, A. (2001). The myth of the ethical consumer—do ethics matter in purchase behaviour? *Journal of consumer marketing, 18*(7), 560–578.

Crane, F. G., & Clarke, T. K. (1994). *Consumer behaviour in Canada: Theory and practice*. Toronto, ON: Dryden Press.

D'Arpizio, C. (2016, May 24). *The global personal luxury goods market in 2016 will mirror last year's low single-digit real growth, even as geopolitical turmoil and luxury brands' emerging strategies reshuffle internal market dynamics*. Bain and Company. Retrieved from http://www.bain.com/about/press/press-releases/spring-luxury-update-2016.aspx.

D'Arpizio, C., Levato, F., Zito, D., & de Montgolfier, J. (2015). *Luxury goods worldwide market study (Fall–Winter 2015). A time to act: How luxury brands can rebuild to win*. Bain and Company. Retrieved from http://www.bain.com/Images/BAIN_REPORT_Global_Luxury_2015.pdf.

Daneshkhu, S. (2013, May 26). Luxury bag the best skins. *Financial Times*. Retrieved from https://www.ft.com/content/ed5824f8-c133-11e2-9767-00144feab7de.

Davies, I., Lee, A., & Ahonkhai, Z. (2012). Do consumers care about ethical-luxury? *Journal of Business Ethics, 106*(1), 37–51.

Diderich, J. (2017, April 4). *H and M ups its green game*. WWD. Retrieved from http://wwd.com/fashion-news/textiles/hm-pledges-use-only-sustainable-materials-10857323/.

East, R., Hammond, K., & Lomax, W. (2008). Measuring the impact of positive and negative word of mouth on brand purchase probability. *International Journal of Research in Marketing, 25*(3), 215–224.

Ellen, P. S., Webb, D. J., & Mohr, L. A. (2006). Building corporate associations: Consumer attributions for corporate socially responsible programs. *Journal of the Academy of Marketing Science, 34*(2), 147–157.

Exposed: Crocodiles and alligators factory-farmed for Hermès 'Luxury' goods. (n.d.). Retrieved from http://investigations.peta.org/crocodile-alligator-slaughter-hermes/ .

Fereday, J., & Muir-Cochrane, E. (2006). Demonstrating rigor using thematic analysis: A hybrid approach of inductive and deductive coding and theme development. *International Journal of Qualitative Methods, 5*(1), 80–92.

Fitzgerald, A. J. (2010). A social history of the slaughterhouse: From inception to contemporary implications. *Human Ecology Review, 17*(1), 58–69.

Folkes, V. (1984). Consumer reactions to product failure: An attributional approach. *Journal of Consumer Research, 10,* 398–409.

Folkman, S., & Lazarus, R. S. (1985). If it changes it must be a process: Study of emotion and coping during three stages of a college examination. *Journal of Personality and Social Psychology: Personality Processes and Individual Differences, 48*(1), 150–170.

Hennigs, N., Karampournioti, E., & Wiedmann, K. P. (2016). Do as you would be done by: The importance of animal welfare in the global beauty care industry. In S. S. Muthu & M. A. Gardetti (Eds.), *Green fashion* (Vol. 1, pp. 109–125). Berlin: Springer Science+Business Media.

Hess, R. L., Ganesan, S., & Klein, N. M. (2007). Interactional service failures in a pseudorelationship: The role of organizational attributions. *Journal of Retailing, 83*(1), 79–95.

Hussin, M., Frazier, S., & Thompson, J. K. (2011). Fat stigmatization on YouTube: A content analysis. *Body Image, 8*(1), 90–92.

Kelley, H. H., & Michela, J. L. (1980). Attribution theory and research. *Annual Review of Psychology, 31*(1), 457–501.

Klein, J., & Dawar, N. (2004). Corporate social responsibility and consumers' attributions and brand evaluations in a product–harm crisis. *International Journal of Research in Marketing, 21*(3), 203–217.

Kotler, P., Kartajaya, H., & Setiawan, I. (2010). *Marketing 3.0: From products to customers to the human spirit*. Hoboken, NJ: Wiley.

Laczniak, R. N., DeCarlo, T. E., & Ramaswami, S. N. (2001). Consumers' responses to negative word-of-mouth communication: An attribution theory perspective. *Journal of consumer Psychology, 11*(1), 57–73.

Lawrence, B., Fournier, S., & Brunel, F. (2013). When companies don't make the ad: A multimethod inquiry into the differential effectiveness of consumer-generated advertising. *Journal of Advertising, 42*(4), 292–307.

Low, W., & Davenport, E. (2007). To boldly go… Exploring ethical spaces to re-politicise ethical consumption and fair trade. *Journal of Consumer Behaviour, 6*(5), 336–348.

Luce, M. (1998). Choosing to avoid: Coping with negatively emotion-laden consumer decisions. *Journal of Consumer Research, 24*(4), 409–433.

McKevitt, J. (2017, March 8). Uniqlo strives for higher sustainability model after controversy. *Supply Chain Dive*. Retrieved from http://www.supplychaindive.com/news/uniqlo-ethical-sourcing-sustainability-report/437601/.

Molderez, I., & De Landtsheer, P. (2015). Sustainable fashion and animal welfare: Non-violence as a business strategy. In *Business, ethics and peace* (pp. 351–370). Bingley: Emerald Group Publishing Limited.

Moraes, C., Carrigan, M., & Szmigin, I. (2012). The coherence of inconsistencies: Attitude–behaviour gaps and new consumption communities. *Journal of Marketing Management, 28* (1–2), 103–128.

Nyilasy, G., Gangadharbatla, H., & Paladino, A. (2014). Perceived greenwashing: The interactive effects of green advertising and corporate environmental performance on consumer reactions. *Journal of Business Ethics, 125*(4), 693–707.

Otter, C. (2008). Civilizing slaughter: The development of the British Public Abattoir, 1850–1910. In P. Young Lee (Ed.), *Meat, modernity and the rise of the slaughterhouse* (pp. 89–106). Durham: University of New Hampshire Press.

Patton, M. Q. (1990). *Qualitative evaluation and research methods*. California: SAGE Publications, inc.

PETA. (n.d.). Retrieved from http://investigations.peta.org/crocodile-alligator-slaughter-hermes/.

Phau, I., Teah, M., & Chuah, J. (2015). Consumer attitudes towards luxury fashion apparel made in sweatshops. *Journal of Fashion Marketing and Management, 19*(2), 169–187.

Rice, P. L., & Ezzy, D. (1999). *Qualitative research methods: A health focus* (Vol. 720). Melbourne.

Sterlacci, F. (1997). *Leather apparel design*. New York: Delmar Publishers.

The Hartman Group. (2015). Sustainability 2015: Transparency. Retrieved from http://store.hartman-group.com/content/sustainability-2015-overview.pdf, 19 March 2018.

Thomas, D. (2015). The luxury sector now focusing on a sustainable future. *New York Times*. Retrieved from https://www.nytimes.com/2015/12/02/fashion/luxury-brands-focusing-on-a-sustainable-future.html?_r=0.

Vigneron, F., & Johnson, L. W. (2004). Measuring perceptions of brand luxury. *The Journal of Brand Management, 11*(6), 484–506.

Weiner, B. (2000). Attributional thoughts about consumer behavior. *Journal of Consumer Research, 27*(3), 382–387.

Wendlandts, A. (2013, August 23). Big brands race to secure luxury supplies from reptiles to roses. *The Reuters*. Retrieved from http://www.reuters.com/article/us-luxury-supplychain-idUSBRE97M0CT20130823.

Wiedmann, K.-P., Hennigs, N., & Siebels, A. (2007). Measuring consumers' luxury value perception: A cross-cultural framework. *Academy of Marketing Science Review, 2007*.

Williams, N. M. (2008). Affected ignorance and animal suffering: Why our failure to debate factory farming puts us at moral risk. *Journal of Agricultural and Environmental Ethics, 21*(4), 371–384.

Winston, A. (2017, January 11). An inside view of how LVMH makes luxury more sustainable. *Harvard Business Review*. Retrieved from https://hbr.org/2017/01/an-inside-view-of-how-lvmh-makes-luxury-more-sustainable/.

YouTube. (n.d.). Retrieved from https://www.youtube.com/yt/press/statistics.html.
Yoon, S. (2013). Do negative consumption experiences hurt manufacturers or retailers? The influence of reasoning style on consumer blame attributions and purchase intention. *Psychology and Marketing, 30*(7), 555–565.
Yoon, Y., Gürhan-Canli, Z., & Schwarz, N. (2006). The effect of corporate social responsibility (CSR) activities on companies with bad reputations. *Journal of Consumer Psychology, 16*(4), 377–390.

Chapter 6
Mining Social Media Data to Discover Topics of Sustainability: The Case of Luxury Cosmetics Brands and Animal Testing

Chao Min, Eunmi Lee and Li Zhao

Abstract Animal experiments have been considered necessary procedures for safety verification and effectiveness validation in developing the products that directly affect the human body, such as medicines or cosmetics (Baumans, 2010; Hajar, 2011). For ending animal testing, social media can be a useful and effective tool for those opposed to animal testing and has been shown to produce measurable results (Wilkinson, 2014). The current research regarding sustainability and animal testing hasn't sufficiently taken advantage of the large-scale data set available online. By applying data-mining-based social network analysis, this study used the French cosmetics company NARS as an example to examine how public awareness and reaction to animal experiments is spread on social media. To quantify and identify the online discussion of Instagram and Twitter users across time, we analyzed two networks of hashtags connected through user posts. To generate the nodes, we first crawled all posts containing #animaltesting within four months for Instagram, one week for Twitter. In both networks, nodes are hashtags created by users when they publish posts on certain events. The findings will be useful for cosmetics companies, lawmakers, and animal advocacy organizations in understanding the network and information flow on social media and, in turn, know what information should be posted on social media to engage social media users and build positive brand reputation.

Keywords Animal testing · Data mining · Network · Sustainability

C. Min
School of Information Management, Nanjing University,
163 Xianlin Avenue, Nanjing 210023, Jiangsu, China
e-mail: marlonmassine@yeah.net

E. Lee · L. Zhao (✉)
Textile and Apparel Management, University of Missouri,
137 Stanley Hall, Columbia, MO 65211, USA
e-mail: zhaol1@missouri.edu

E. Lee
e-mail: eunmilee@mail.missouri.edu

© Springer Nature Singapore Pte Ltd. 2018

C. K. Y. Lo and J. Ha-Brookshire (eds.), *Sustainability in Luxury Fashion Business*,
Springer Series in Fashion Business, https://doi.org/10.1007/978-981-10-8878-0_6

Introduction

Animal experiments have been considered necessary procedures for safety verification and effectiveness validation in developing the products that directly affect the human body, such as medicines or cosmetics (Baumans, 2010; Hajar, 2011). Millions of animals are used in experiments worldwide each year, and the number of animals used in research has increased with technology advances in the area of medicine and cosmetics (Doke & Dhawale, 2015). As a result, interest in animal welfare that minimizes human's stress on animals and the realization of psychological happiness of animals has increased greatly, especially in developed countries including European nations and the USA (Dawkins, 2012). Recent studies show that Americans' verbal opposition to animal testing has grown significantly since 2001 among people of every gender, age group, and political affiliation (Bruner, 2014). The results are encouraging to animal advocacy organizations and demonstrate how much public opinion regarding animal testing has changed in a fairly short amount of time.

A possible reason is that people have more exposure to information about the cruelty that animals endure in laboratories, and the alternatives that are available (Bruner, 2014). One important channel through which the public can receive information is social media. For ending animal testing, social media can be a useful and effective tool for those opposed to animal testing and has been shown to produce measurable results (Wilkinson, 2014).

In recent years, social media has developed rapidly and has drastically transformed the way in which people communicate and collect information. Social media has become ubiquitous and plays an increasingly critical role in the exchange of information and opinions among the public. As a result, a large amount of user-generated content is available on social media sites such as Facebook, Twitter, and Instagram. Sustainability topics are also under active discussion on social media, and the public and consumers have more access to information on companies' efforts and practices to achieve sustainability (Goswami & Ha-Brookshire, 2015). However, the current research regarding sustainability and animal testing has not sufficiently taken advantage of the large-scale dataset available online. Only a few studies have focused on animal testing and social media, and most of those utilized surveys and interviews to discover how social media could promote sustainable practices (Scholtz, Burger, & Zita, 2016).

A large and diverse audience on the Internet expresses and shares opinions and provides feedback to other users, including media, businesses, and government. Thus, it is often necessary for businesses and policy makers to collect, monitor, and analyze user-generated data on social media sites. Accordingly, these large-scale datasets can be used to glean and identify the needs of the public and consumers and can generate meaningful insights into businesses and policy makers with respect to sustainability topics such as animal testing. Based on the feedback and ideas from social media users, specific actionable areas in which businesses or policy makers are leading and lagging can be found and this insight can further improve their performance in a wide array of fields. In this light, this study aims to (1) investigate social media activities regarding animal testing through a large amount of user-generated content and (2) identify key influencers and major communities discussing animal testing by utilizing data mining-based social network analysis.

Review of Literature

Current Policies Concerning Sustainability and Animal Testing

Sustainable development has been defined as "development that meets the needs of the present without compromising the ability of future generations to meet their own needs" (Brundtland Commission, 1987). Sustainability offers "win–win opportunities" to reconcile "environmental protection and smart economic growth" (European Union, 2006). In this light, animal protection represents such an opportunity and becomes one of the key objectives of sustainable development strategy (Keeling, 2005). In the 1950s, a so-called 3Rs principle of reduction, refinement, and replacement (Russel & Burch, 1959) was proposed and then expanded to promote the welfare of animals used in experiments (Bouhifd et al., 2012). More and more efforts have been developed to minimize or prohibit animal testing mainly in developed countries such as the European Union (EU) member states and the USA. For example, the EU adopted the regulation Registration, Evaluation, and Authorization of Chemicals (REACH) in 2007 to improve human health and environmental protection (European Chemicals Agency, 2017). Further, the EU has banned all animal testing for cosmetics in Europe since March 2013 (European Commission, 2013). These policies and regulations have led to pressure on the industry and promotion of the development of alternative methods to replace animal testing conducted to verify product safety (Basketter et al., 2012).

The cosmetics industry, in which testing on animals to make sure they are safe for consumers was widespread, is considered the area most impacted by these efforts (Long, 2016). Many cosmetics brands, such as L'Occitane, NARS, Estee Lauder, Benefit, and others, are not "cruelty-free" and still use animal testing (Chitrakorn, 2016). Despite the wide availability of alternatives, countless animals are still subjected to hazardous tests around the world due to industry inertia and bureaucracy. However, the EU, the world's largest cosmetics and personal care market, became one of the first regions in the world to ban animal testing for cosmetics (European Commission, 2013). A snowball effect can be observed as more and more countries introduce laws and proposals aimed at eliminating animal testing. In the USA, a bill called the Humane Cosmetics Act (H.R. 4148), which prohibits animal experiments in phases during the development of cosmetics, was initiated in 2014, but until recently, no progress had been made toward its passage. On June 6, 2017, the bill, which outlaws the development, sales, and transportation of animals for the use of animal experiments, entered the first stage of the legislative process (United States Congress, 2017). Figure 6.1 shows the historical paths and

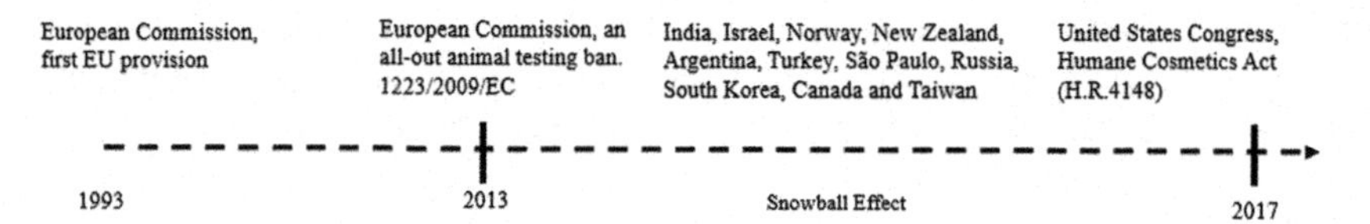

Fig. 6.1 Historical paths of animal testing policies in major countries related to cosmetics industry

the current status of animal testing policies in major countries in the world, especially in the cosmetics industry.

The biggest barrier to progress lies with China, where the cosmetics industry has shown rapid economic development (Lopaciuk & Loboda, 2013). China is a giant cosmetics market that cannot be ignored by multinational cosmetics companies, and many cosmetics and healthcare research and development centers there conduct experiments on various animals (Lu, Bayne, & Wang, 2013). Under the law amended on June 30, 2014, China issued the legal requirement that makeup products, perfumes, and skin, hair and nail care products manufactured and sold in China must be first tested on animals (Human Society International, 2017). The People for the Ethical Treatment of Animals (PETA) estimated that about 300,000 animals were sacrificed due to the mandatory Chinese animal testing enforcement since 2014 (Yan, 2017). China has been challenged by international animal rights organizations and NGOs over animal welfare and protection, and groups such as PETA are actively holding campaigns against animal experimentation and exports (Bayne, Ramachandra, Rivera, & Wang, 2015; Lu et al., 2013). The Chinese government has also been under the pressure of trade sanctions due to animal testing since joining the World Trade Organization (WTO) (Lu et al., 2013). In particular, PETA has been working to persuade the Chinese government to accept test results through alternative methods other than animal testing, and the Chinese government has stated that it will accept data from a non-animal test method for safety evaluations of cosmetics (PETA, 2012; PETA, 2016). However, the current laws and regulations of the Chinese Food and Drug Administration (CFDA) are still demanding animal testing results (Human Society International, 2017). With these small changes taking place, advocates for an animal welfare group are anticipating more advances under the policy on animal testing in China.

Social Media Influence on Sustainability and Animal Testing

The trend of public pressure is toward ending animal testing. Individuals, policy makers, and cosmetics companies will need to engage in a wide range of non-animal testing practices to address problems and challenges. As more and more consumers become aware of the realities of animal testing, dealing with public relations will be a nightmare for cosmetics companies (Chitrakorn, 2016; D'Souza, Taghian, Lamb, & Peretiatkos, 2006; Ramli, 2016). The emergence of social media has created an opportunity for organizations such as companies and governments to manage their reputations and maintain relationships with the public (Lovejoy & Saxton, 2012; Wright & Hinson, 2011). Previous studies have found how the ubiquity of social media affects government interests, and policy accordingly reacts to it (Shirky, 2011). Therefore, social media can bring various stakeholders of the cosmetics industry together onto a new playing field.

Social media is widely used to acquire and share new knowledge and build relationships and is a useful tool for spreading campaigns around the world through the reach of online communication without cost and distance constraints (Hou & Lampe, 2015). According to the Social Media Industry Report (Stelzner, 2013), corporate marketers consider social media channels to be an important component of their marketing strategy and many nonprofit organizations in the USA believe that social media is an innovative communication means that will replace traditional Web sites (Nonprofit Technology Network, 2011). Using social media, positive messages such as animal protection campaigns are effectively conveyed, but it is also a vehicle for spreading information about negative events and forging public opinion. When a big news event occurs, social media spreads and propagates the information through social networks to people who are not aware of the news. Since July 28, 2017, when NARS, a French cosmetics company, stated on its Instagram account that animal testing is necessary to sell their products in China (NARS, 2017), there have been 250,000 e-petitions calling on the company to stop selling its products in China (Salemme, 2017). It is apparent that the use of social media among the younger generation is relatively high and social media is a powerful tool to deliver messages.

A few studies have researched how the public becomes aware of and reacts to animal experiments and the resulting shifting attitudes on social media (Goodman, Borch, & Cherry, 2012). The Gallup organization conducted an annual "Values and Beliefs" survey of 1000 American adults from 2001 to 2013 (PETA, 2014). Results showed that since 2001, there has been a yearly increase in the number of respondents being of the opinion that animal experiments are morally wrong. The results were disparate according to genders, respondents' age groups, and education levels. With respect to age groups, the younger generation from 18 to 29 years of age (54%) was more negative (opposed to testing) than those in their 30 s (about 33%) (PETA 2014). Interestingly, the younger generation's opposition to animal experiments increased to 54% in 2013, compared to 31% in 2001. This trend is due to the increased use of the Internet and social media during this period (PETA, 2014), along with an increase in interest in animal welfare (Morell, 2014). The fact that organizations such as PETA and the Foundation for Biomedical Research, which work on behalf of animal rights and welfare, have a strong presence on Facebook, Instagram, and Twitter indicated that social media has a great influence on the animal testing issue (Grimm, 2014; Wilkinson, 2014). Social media campaigns by animal rights organizations have been successful in raising public awareness of certain issues related to animal experimentation (Ormandy & Schuppli, 2014). This is further illustrated by the large memberships of social media groups with an animal rights or welfare focus (Ormandy & Schuppli, 2014) such as PETA, which has over 5.3 million Facebook likes and 1.04 million Twitter followers (PETA "Home," 2017). It is apparent that social media has a great influence on forming public opinion related to animal testing.

Nonetheless, the role of social media in driving the opposition to animal testing and shaping public debates has not been sufficiently investigated. This study investigated how the issue of animal testing has been communicated, how it is shared, and what kinds of responses related to the topic can be found on social media. Specifically, as described in the next section, one event related to animal testing on social media was selected and a novel approach—data mining-based social network analysis—was used to analyze social media data.

Context and Approach

Research Context

This study focused on the two dominant social network sites worldwide: Twitter and Instagram. Twitter, which was founded in 2006, has around 330 million monthly active users as of the third quarter of 2017 (Statista, 2017a, b). It allows its users to post short messages limited to 140 characters and upload photographs or short videos via the Web or a mobile phone (Statista, 2017a, b). As one of the most popular social media platforms, Twitter enables its members to share and discover topics of their choice in real time (Statista, 2017a, b). One unique feature of Twitter is known as a "retweet," which enables its users to forward a tweet to their followers. This retweet function facilitates rapid dissemination of information to a larger public since retweets often reach beyond the original tweet's followers (Kwak, Lee, Park, & Moon, 2010). Another characteristic of Twitter is the use of hashtags and the reply function. A hashtag is a convention among Twitter users to set a thread of discussion by prefixing a word with a "#" character. This hashtag allows users to identify and emphasize their topics of interests and effectively target the intended audience (Thelwall, Buckley, & Paltoglou, 2011). The reply symbol, which is indicated as "@," allows Twitter users to post their messages to another Twitter user, thus facilitating effective discussions and engagement of larger audiences (Kwak et al., 2010). The @ sign and the hashtags are effective strategies used by Twitter users in order to relate one tweet to another in regard to a certain topic in real time. Therefore, messages and opinions are able to rapidly reach a wider audience (Thelwall et al., 2011).

Another form of communication, Instagram users can easily share their updates by taking photographs and short videos and adding hashtags to link the photographs and videos up to other content on Instagram featuring the same subject or overall topic (Sheldon & Bryant, 2016). Through the platform, other users can follow, view, like, and comment on these posts. Additionally, since it was acquired by Facebook in 2012 (Luckerson, 2016), Instagram has become one of the most popular social media platforms worldwide by benefiting from its association with Facebook, which allows Instagram content to be posted on Facebook

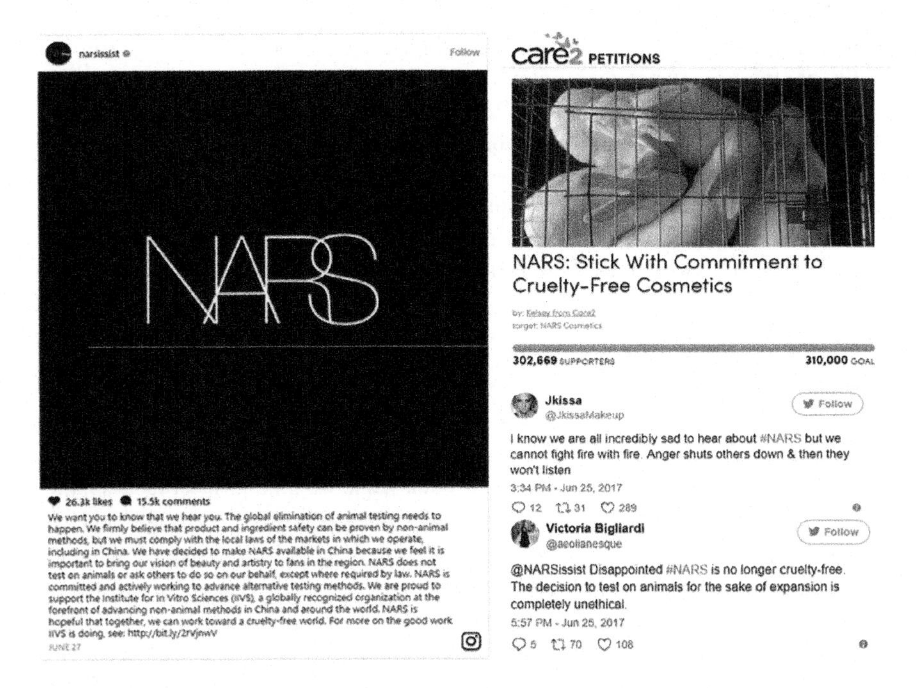

Fig. 6.2 NARS's statement about animal testing on Instagram, Care2 campaign, and consumers' opinions on Twitter. Captured on Oct 27th, 2017

simultaneously (Kawasaki & Fitzpatrick, 2014). It has 800 million monthly active users as of September 2017 (Statista, 2017a, b). Instagram users include both individuals and businesses, and their posts are generally annotated with hashtags, short texts, or keywords to be searchable.

Along with its enormous user base and powerful dissemination with hashtags, Instagram is a valuable social media channel for cosmetics brands to communicate with consumers. More than one million monthly cosmetics commercials were active on Instagram as of March 2017 (Statista, 2017a, b). Moreover, Instagram is powerful among teenagers and young millennials as more than half of the USA. Instagram user base is between the age of 18 and 29. Considering that millennial women ages 18–34 are the dominant consumers of cosmetics products (McCarthy, 2016), it is apparent that Instagram messages, topics, and agendas with respect to cosmetics is highly important to the cosmetics industry. Therefore, one recent event related to animal testing in the cosmetics industry was selected in this study to discover the influence of social media on forming public opinion about animal testing—NARS and #animaltesting. Twitter data was collected accordingly to determine whether people show different opinions on these two social media platforms.

NARS, a luxury makeup brand, recently decided to discontinue its cruelty-free status and conduct testing on animals in order to enter the Chinese market (Saltzman, 2017). It was a disappointing decision for many fans who love the makeup brand. On June 27, 2017, NARS posted a statement on Instagram to explain the circumstances and the company's vision about animal testing. By October 27, 2017, within four months of this post, more than 15,500 comments were made. In addition, a new campaign by Care2 is calling for NARS to stop animal testing that has more than 300,000 signatures on an e-petition. The hashtag #animaltesting is used to join the discussion, and it has been promoted to attract more individual users and become a "trending" topic. The analysis of Instagram posts and tweets with the hashtag #animaltesting in this study revealed what people talk about and feel regarding animal testing. Figure 6.2 shows relevant information about this event.

Research Approach

Data mining-based social network analysis was used in this study. Web sources such as Twitter, Instagram, and Facebook have been commonly used in academic research in recent years (Sheble, Brennan, & Wildemuth, 2016). A wide range of techniques has been used in this research, and techniques for social network analysis using such data are still being developed (Son, 2016). Data mining-based social network analysis was used in this study to investigate the pattern of discussion regarding animal testing on Instagram. Social network analysis (SNA) has been used as an effective tool to understand the social organization of groups based on the associations or interactions between individuals (Wasserman & Faust, 1994). Additionally, social network analysis enables researchers to quantify and interpret the relationships among social entities that may be single individuals, groups of individuals, communities, or organizations (Cronin, 2016).

A social network is a social structure of people related (directly or indirectly) to each other through a common relation or interest (Liu, Sidhu, Beacom, & Valente, 2017). Computer-supported interpersonal communication, such as social media, has changed the way people connect with each other (Wellman, 2001). Computer networks are inherently social networks, linking people, organizations, and knowledge. With the massive data on the Internet, large-scale data sets can be obtained and utilized to understand people's behavior in the digital world. Researchers in computer science have proposed data mining-based social network analysis to describe their experiences in climbing the obstacles of informational challenges (Zuber, 2014). Data mining refers to the process of automated information extraction using as input a variety of complex or unstructured data sources (Feldman & Sanger, 2007). The data embedded into this social network is measured by databases and statistically prepares related messages to categorize the thoughts and people's behavior (Srivastava, Ahmad, Pathak, & Hsu, 2008). Data mining-based social network analysis would be a valuable approach for us to explain complex data sources and the social structure of social media users.

Network exchange theory, network flow theory, small world theory, and the strength of weak ties theory are often used to provide theoretical foundations of social network analysis (Borgatti & Lopez-Kidwell, 2015). The strength of weak ties claims that two people under strong ties tend to have overlapped social lives and have ties with a third person in common (Granovetter, 1973). Additionally, the two people and the third person also tend to be similar in lifestyle and tastes (Granovetter, 1973). By contrast, a person who has loose ties to people and do not have common third persons can be a source of novel ideas and information (Granovetter, 1973). The theory of strength of weak ties has been applied to explain information flow and the ties within social networks. Borgatti and Lopez-Kidwell (2015) argue that the weak ties between persons in different social networks provide powerful strengths in that it has the potential of providing novel information.

Combined with mathematics methodologies, computer science and social science have together enhanced the foundation and presence of social network analysis and visualization as an interdisciplinary field (Bruns, 2012). In particular, researchers employing these methods have shown high interest in online social networks since the analysis of the online networks offers immense insights into researchers through the interleaving of human interactions (Zuber, 2014). Moreover, the abundance of readily accessible and apparently objective data on online social networks has attracted researchers (Bruns, 2012). It is easy and convenient to obtain information on who, what, when, how, where, and why from online social media platforms (Bruns, 2012).

Social network analysis (SNA) is often characterized as being visualized through nodes, ties, links, or edges (D'Andrea et al., 2010). While nodes indicate individual actors, people, or things within social networks, ties and links represent relationships or interactions that connect each node (D'Andrea et al., 2010). Visual representation of social networks analysis is important for delivering and interpreting network data and the result of the analysis (Freeman, 2000). Network visualization has the potential to provide insight into relationships among individual nodes and network structure and provide a much more abundant representation (Cronin, 2016). Numerous methods have been developed to visualize social network analysis (Caschera, Ferri, & Grifoni, 2008). Exploration of the data is implemented through displaying nodes and ties in a wide array of layouts and assigning diverse colors, size, and other advanced properties to nodes and links. Visual representations of networks may be a powerful tool to efficiently deliver complex information, but special attention is required to interpret nodes and graphics from visual displays.

Method

Instagram posts with the hashtag #animaltesting were crawled and limited dates from June 27, 2017, to October 27, 2017, four months since NARS posted a statement on Instagram to explain the circumstances of its decision and its vision about animal testing. Researchers obtained 3481 posts, among which 9029 hashtags were extracted and 22,617 edges between these hashtags were constructed based on hashtag co-occurrence in an Instagram post. In addition, to compare different social media platforms, tweets with the hashtag #animaltesting were crawled and limited dates from June 27, 2017, to July 4, 2017, one week from the NARS statement. Researchers obtained 3132 tweets, among which 251 hashtags were extracted and 251 edges between these hashtags were constructed based on hashtag co-occurrence in a tweet.

To quantify and identify the online discussion of Instagram and Twitter users across time, we analyzed two networks of hashtags connected through user posts. In both networks, nodes are hashtags created by users when they publish posts on certain events. To generate the nodes, we first crawled all posts containing #animaltesting within four months for Instagram, one week for Twitter. We then extracted all hashtags, which are naturally good representations of topics being discussed, from the Instagram and Twitter data. These became the initial nodes for the networks. To create edges between nodes, we look to the co-occurrence of hashtags in the same Instagram post or tweet, connecting two nodes if they occurred in the same post.

After the dataset was prepared, Gephi was used to analyze and visualize Instagram and Twitter data. Force Atlas layout algorithm was applied for Instagram network visualization. Fruchterman Reingold layout algorithm was employed for Twitter network visualization. With different algorithms, we aim to demonstrate different layout and visualization effects. Degree centrality, betweenness centrality, closeness centrality, and K-core were used as key indicators to measure the importance of a node in the network (Kitsak et al., 2010). As there can be hundreds of nodes and thousands of edges in the networks, we applied filters for some conditions (e.g., limiting node degree, K-core, or tag frequency) to present the core structure of topic discussion networks. We used Gephi's algorithms to roughly categorize the nodes into different clusters where hashtags in the same cluster are viewed to discuss related topics. For a better visualization effect, the nodes and edges are colored according to the groups that the nodes belong to.

Results and Discussion

Instagram Network with #Animal Testing

Figure 6.3 provides a visualized #animaltesting network on Instagram. Animal testing was shown as the central node in the Instagram network. According to

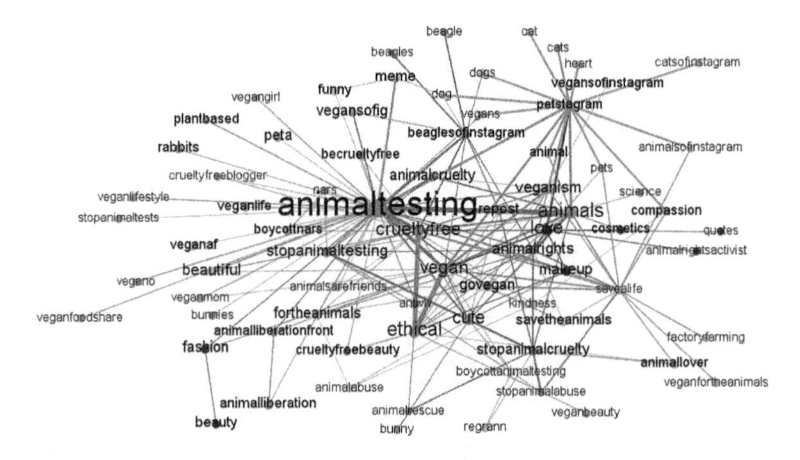

Fig. 6.3 Visualized #animaltesting network on Instagram

betweenness centrality, which indicates how often it appears between any two random nodes in the network (Freeman, 1977; Brandes, 2001), several nodes appeared as key junctions for communication within the network. These nodes were more influential as they were with the higher betweenness centrality. These nodes include #petstagram, #savealife, #vegan, #crueltyfree, #ethical, and #beagles of Instagram. Lots of Instagram users care about animal testing-related information because of their pets. Pet owners love to share photographs of their pets on social media. #petstagram and #beaglesofInstagram have turned many pets to social media stars. Pet owners love their pets and often think of their dogs and cats as members of the family (Kaplan, 2017). Empathy motivates them to promote the campaign against animal testing.

The network from Instagram regarding animal testing demonstrated a wide range of sub-themes as well. Distinct hashtag communities were present within the network classified by different colors. Pink color with the hashtag #animaltesting as the central topic emerged as the largest community in the network. Other topics with this community include #boycott NARS, #stopanimaltesting, #animalcruelty, #becrueltyfree, #vegansofig, #veganlife, #rabbits, #peta, #beaglesofinstagram. From the texts, it can be assumed that the communities that belong to the pink color are the most proactive in the opposition to animal testing and have a high interest in practicing vegan life. Rabbits and beagles are commonly used animals for experiments and PETA is an animal rights organization with the slogan "Animals are not ours to eat, wear, experiment on, use for entertainment, or abuse in any other way" (PETA, 2017). Additionally, pink color community indicated a great interest in vegan life with the hashtags #veganfoodshare, #veganmon, #vegano, #veganaf, #veganlifestyle, #veganlife, #vegansofig, #vegangirls, and others.

The second largest community in brown color with the main hashtag #petstagram indicated affection for animals and animal welfare. Other topics in this network include #dog, #cat, #heart, #pets, #animalsofinstagram, #catsofinstagram. It can be assumed that they own dogs or cats, so they want to share their love for animals and interact with other Instagram users who have the same concerns and interests. The community with blue color texts mainly showed the concerns about animal testing itself. Its hashtags include #crueltyfree, #ethical, #vegan. Green color texts indicated high relevance to brown color texts in that they care about animal love and rights. Its hashtags include #animals, #animalrights, # love, #savealife, #science, #animalcruelty.

Notably, #makeup, #cosmetics, #beauty, and #fashion were shown in the network with gray color community. As NARS is a well-known cosmetics brand, Instagram users wish these brands can work toward a cruelty-free world. Hashtags #makeup and #fashion have a relatively higher degree for their betweenness centrality. It means they are important local hubs for certain communities. However, they are not serving as important junctions for meaning circulation within the whole network. That is, there is a group of people who might be specifically interested in cosmetics brands and animal testing. However, beauty and fashion brands have not taken enough responsibilities to eliminate animal testing. Table 6.1 shows the top 15 nodes in the network and their betweenness centrality, top 15 nodes with degree information, and top 15 co-occurrence times on Instagram with #animal testing (Fig. 6.3).

Table 6.1 Instagram top 15 nodes with degree information, top 15 nodes in the network and their betweenness centrality, and top 15 co-occurrence times

Top 15 nodes with degree		Top 15 betweenness centrality		Top 15 co-occurrence times		
Label	Degree	Label	Betweenness centrality	Source	Target	Weight
Animal testing	1575	Animal testing	2912.207	Animals	Animal testing	100
Animals	575	Petstagram	850.344	Ethical	Vegan	94
Ethical	552	Save a life	664.256	Ethical	Cruelty Free	86
Love	512	Vegan	506.644	Animal testing	Cruelty Free	68
Cruelty free	436	Cruelty free	490.745	Animal testing	Vegan	66
Vegan	417	Animals	382.624	Petstagram	Animal testing	50
Cute	357	Ethical	352.574	Petstagram	Animals	50
Animal rights	292	Beagles of instagram	309.002	Petstagram	Animal	49
Veganism	285	Love	281.084	Petstagram	Love	48
Makeup	281	Meme	222.951	Petstagram	Dogs	47
Stop animal testing	272	Veganism	194.498	Petstagram	Dog	47
Beautiful	258	Go vegan	172.985	Petstagram	Pets	47
Art	251	Animal rights	167.774	Meme	Animal testing	46
Fashion	208	Cute	132.844	Petstagram	Cats	46
Peta	206	Kindness	132.605	Petstagram	Compassion	46

Twitter Network with #Animal Testing

Similar to the Instagram network, #animal testing was shown as the central node in the twitter network. The hashtag #love, #cruelty free, #morals, #3Dpring emerged as the most influential nodes within the network. Interestingly, Twitter users focused more on non-animal alternatives, such as 3D printing. Using 3D-printed skin to evaluate cosmetics is a hot topic in the beauty industry (Ashton et al., 2014). 3D skin innovation is viewed as an alternative to animal testing (Ashton et al., 2014). Also, #worldwide, #canada, and #fukushima were present in the network. According to the latest data of Twitter user demographics, 79% of Twitter accounts are based outside the USA (Aslam, 2017). Compared to Instagram, animal testing attracts attention worldwide on Twitter. Several distinct communities were categorized by different colors in the network. Most communities are similar to Instagram. However, it shows more variety of topics and interest. Not surprisingly,

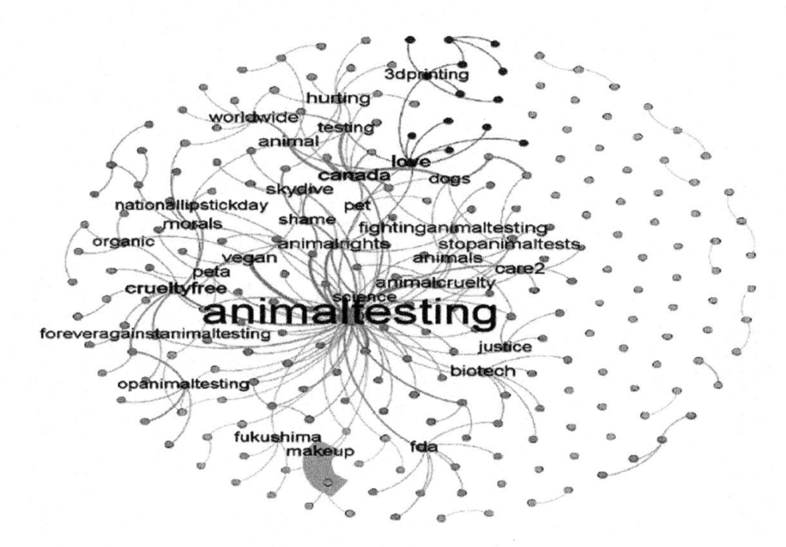

Fig. 6.4 Visualized #animaltesting network on Twitter

#makeup and #national lip stick day indicated that animal testing issues are associated with the beauty industry. Especially, according to co-occurrence times, when Twitter users tag #animaltesting, they often tag #NARS at the same time, which indicates the brand image of NARS is now negatively associated with animal testing. Table 6.2 shows the top 15 nodes in the network and their betweenness centrality, top 15 nodes with degree information, and top 15 co-occurrence times on Twitter with #animaltesting (Fig. 6.4).

Conclusions and Implications

Animal experiments have become a worldwide topic that attracts attention from industry professionals, consumers, policy makers, and scholars (Hou & Lampe, 2015). Social media can be a useful tool to educate consumers, build relationships, and promote sustainability. This study used the French cosmetics company NARS as an example to examine how public awareness and reaction to animal experiments is spread on social media. Social network analysis was used to study interactions regarding #animaltesting between social media users on both Instagram and Twitter. Social network analysis provides techniques to analyze the structure of a network and identify influential nodes in online communities. Four months of data after the NARS event were tracked on Instagram while one week of data was tracked on Twitter. Social media users on both Instagram and Twitter showed high interest on #vegan, #crueltyfree, #ethical, and #love. Several influential nodes revealed that the beauty industry is highly related to animal testing issues.

Table 2 Twitter top 15 nodes with degree information, top 15 nodes in the network and their betweenness centrality, and top 15 co-occurrence times

Top 15 nodes with degree		Top 15 betweenness centrality		Top 15 co-occurrence times		
Label	Degree	Label	Betweenness centrality	Source	Target	Weight
Animal Testing	66	Animal testing	13864.799	Makeup on animals	Makeup	35
Cruelty free	12	Love	2931.284	Canada	Save the dogs	6
Canada	12	Canada	1864.230	Animal testing	Vegan	5
Love	10	Cruelty Free	1845.558	Animal testing	Animal cruelty	4
Hurting	9	Animal	1629.094	Animal testing	Boycott	4
Animal rights	8	Morals	1515.994	Animal testing	NARS	4
Care2	8	3D printing	1240	Canada	Testing	4
Biotech	8	FDA	1153.833	Canada	Animal	4
FDA	8	Hurting	1007.682	Canada	Dogs	4
Worldwide	8	Biotech	981.1667	Animal testing	Animal rights	3
Animal	7	Worldwide	858.856	Animal testing	Cruelty free	3
Animal cruelty	7	National lipstick day	751.330	Canada	Share	3
Peta	7	Organic	726.833	Domains	Cats	3
Stop animal tests	7	Care2	724.5	Domains	Dogs	3
Animals	6	Go vegan	715	Animal research	Animal testing	2

The hashtags #makeup, #fashion, and #beautiful on Instagram and #nationallipstickday and #NARS on Twitter all showed that the NARS event had an enormous impact on attitudes and perceptions toward animal testing on social media. Social media users showed their support of using cruelty-free cosmetics and ending animal testing for beauty products.

Instagram users often link their love of pets to the anti-animal testing mindset and action. However, Twitter users focus on alternatives to animal testing, such as 3D printing and biotech. Twitter reached a variety of audiences and attracted worldwide intention. Interestingly, #Canada emerged on Twitter due to the news that the Cruelty-Free Cosmetics Act (S-214) was passed by the Senate Standing Committee on Social Affairs, Science and Technology in Canada (Graef, 2017).

It shows that Canada could be next to ban animal testing for cosmetics. In addition, some animal advocacy organizations such as PETA and Care2 influenced a great many people on social media. PETA and Care2 have a high number of people liking, sharing, and commenting on their tweets and Instagram pictures. They have led many social media campaigns and try to reach consumers around the world and convince them to avoid cosmetic companies that test their products on animals.

The findings will be useful for cosmetics companies, lawmakers, and animal advocacy organizations in understanding the network and information flow on social media and, in turn, know what information should be posted on social media to engage social media users and build positive brand reputation. In particular, personal and emotional messages can make people feel touched and remind them of their own pets. Then, more people will help oppose animal testing activities. At the same time, logistical responses should be noticed by cosmetics companies, lawmakers, and animal advocacy organizations. Valuable information such as alternatives to animal testing can convince and guide people to support ending animal testing for cosmetics. In addition, key influencers in each community were identified in this study. By targeting key influencers, cosmetics companies, lawmakers, and animal advocacy organizations will be able to spread relevant information more quickly and effectively. Additionally, data mining-based social network analysis has been confirmed as a promising approach to understanding new phenomena in social networking by taking advantage of large-scale datasets available on Internet.

Limitations and Future Research

This study offers future research opportunities. First, even though four months of data were collected on Instagram, this study did not investigate information mobilization. It would be very interesting to see how social media users shift their attitudes and change their opinions over the time. Second, people from different countries may have their own understanding about animal testing. It would be very important to track geographic locations of social media users in order to customize a social media campaign for different needs and focuses. Third, future research may want to focus on China where the cosmetics industry has seen rapid economic development. Since China has its unique social media platforms, understanding how social media users react to animal testing is critical for educating Chinese cosmetics consumers and promoting non-animal testing procedures.

References

Ashton, R., De Wever, B., Fuchs, H. W., Gaca, M., Hill, E., Krul, C., … & Roggen, E. L. (2014). State of the art on alternative methods to animal testing from an industrial point of view: ready for regulation. *Altex*, *31*(3), 357–363.

Aslam, S. (2017). Twitter by the Numbers: Stats, Demographics & Fun Facts. Omnicore. Retrieved August 12, 2017 from https://www.omnicoreagency.com/twitter-statistics/.

Basketter, D. A., Clewell, H., Kimber, I., Rossi, A., Blaauboer, B. J., Burrier, R., et al. (2012). A roadmap for the development of alternative (non-animal) methods for systemic toxicity testing-t4 report. *ALTEX-Alternatives to Animal Experimentation, 29*(1), 3–91.

Baumans, V. (2005). Science-based assessment of animal welfare: Laboratory animals. *Revue Scientifique Et Technique-Office International Des Epizooties, 24*(2), 503–514.

Bayne, K., Ramachandra, G. S., Rivera, E. A., & Wang, J. (2015). The evolution of animal welfare and the 3Rs in Brazil, China, and India. *Journal of the American Association for Laboratory Animal Science, 54*(2), 181–191.

Borgatti, S.P., & Lopez-Kidwell, V. (2015). Network Theory. In J. Scott & P. Carrington (Eds.), *The SAGE Handbook of Social Network Analysis* (pp. 40–54). London, UK: SAGE Publications Ltd.

Bouhifd, M., Bories, G., Casado, J., Coecke, S., Norlén, H., Parissis, N., ... & Whelan, M. P. (2012). Automation of an in vitro cytotoxicity assay used to estimate starting doses in acute oral systemic toxicity tests. *Food and chemical toxicology, 50*(6), 2084–2096.

Brandes, U. (2001). A faster algorithm for betweenness centrality. *Journal of Mathematical Sociology, 25*(2), 163–177.

Brundtland, G. H. (1987). *Report of the world commission on environment and development: Our common future.* United Nations.

Brundtland Commission. (1987). *Our Common Future.* World Commission on Environment and Development. London, UK: Oxford University Press.

Bruner, T. (2014). New study shows growing opposition to animal tests. Eurekalert—The Global Source for Science News. Retrieved February 16, 2014 from https://www.eurekalert.org/pub_releases/2014-02/pfte-nss021314.php.

Bruns, A. (2012). How long is a tweet? Mapping dynamic conversation networks on Twitter using Gawk and Gephi. *Information, Communication & Society, 15*(9), 1323–1351.

Caschera, M. C., Ferri, F., & Grifoni, P. (2008). SIM: A dynamic multidimensional visualization method for social networks. *PsychNology Journal, 6*(3), 291–320.

Chitrakorn, L. (2016). Is the global cosmetics market moving towards a cruelty-free future? Business of Fashion. Retrieved January 13, 2016 from https://www.businessoffashion.com/articles/intelligence/is-the-global-cosmetics-market-moving-towards-a-cruelty-free-future.

Cronin, B. (Ed.). (2016). *Handbook of research methods and applications in heterodox economics.* Cheltenham, UK: Edward Elgar Publishing.

Cruelty Free International. (2017). EU ban on cosmetics testing. Retrieved from https://www.crueltyfreeinternational.org/what-we-do/corporate-partnerships/eu-ban-cosmetics-testing.

D'Andrea, A., Ferri, F., & Grifoni, P. (2010). An overview of methods for virtual social networks analysis. In *Computational social network analysis* (pp. 3–25). London, UK: Springer.

Dawkins, M. (2012). *Animal suffering: The science of animal welfare.* Berlin/Heidelberg, Germany: Springer Science & Business Media.

Doke, S. K., & Dhawale, S. C. (2015). Alternatives to animal testing: A review. *Saudi Pharmaceutical Journal, 23*(3), 223–229.

D'Souza, C., Taghian, M., Lamb, P., & Peretiatkos, R. (2006). Green products and corporate strategy: An empirical investigation. *Society and business review, 1*(2), 144–157.

European Chemicals Agency. (2017). Understanding REACH. Retrieved from https://echa.europa.eu/regulations/reach/understanding-reach.

European Commission. (2013). Communication from the Commission to the European Parliament and the Council on the animal testing and marketing ban and on the state of play in relation to alternative methods in the field of cosmetics. Retrieved from http://ec.europa.eu/consumers/sectors/cosmetics/files/pdf/animal_testing/com_at_2013_en.pdf.

European Commission. (2017). Ban on Animal Testing. Retrieved from https://ec.europa.eu/growth/sectors/cosmetics/animal-testing_en.

European Union. (2006). Renewed EU Sustainable Development Strategy. Retrieved from http://register.consilium.europa.eu/doc/srv?l=EN&f=ST%2010117%202006%20INIT.

Feldman, R., & Sanger, J. (2007). *The text mining handbook: Advanced approaches in analyzing unstructured data*. Cambridge, UK: Cambridge University Press.

Freeman, L. C. (1977). A set of measures of centrality based on betweenness. *Sociometry, 40*(1), 35–41.

Freeman, L. C. (2000). Visualizing social networks. *Journal of social structure, 1*(1), 4.

Gifford, R., & Nilsson, A. (2014). Personal and social factors that influence pro-environmental concern and behaviour: A review. *International Journal of Psychology, 49*(3), 141–157.

Goodman, J. R., Borch, C. A., & Cherry, E. (2012). Mounting opposition to vivisection. *Contexts, 11*(2), 68–69.

Goswami, S., & Ha-Brookshire, J. (2015). From compliance to a growth strategy: Exploring historical transformation of corporate sustainability. *Journal of Global Responsibility, 6*(2), 246–261.

Graef, A. (2017). Canada could be next to ban animal testing for cosmetics. Care2. Retrieved October 11, 2017 from http://www.care2.com/causes/canada-could-be-next-to-ban-animal-testing-for-cosmetics.html.

Granovetter, M. S. (1973). The strength of weak ties. *American Journal of Sociology, 78*(6), 1360–1380.

Grimm, D. (2014). Is Social Media Sourcing Americans on Animal Research? Retrieved February 16, 2014 from http://www.sciencemag.org/news/2014/02/scienceshot-social-media-souring-americans-animal-research.

Hajar, R. (2011). Animal testing and medicine. *Heart Views: The Official Journal of the Gulf Heart Association, 12*(1), 42.

Hou, Y., & Lampe, C. (2015, April). Social media effectiveness for public engagement: Example of small nonprofits. In *Proceedings of the 33rd annual ACM conference on human factors in computing systems* (pp. 3107–3116). New York: ACM.

Human Society International. (2017). China's cosmetics animal testing FAQ. Retrieved August 30, 2017, from http://www.hsi.org/assets/pdfs/bcf_china_cosmetics.pdf.

Kaplan, S. (2017). Dear science: Why do we love our pets. The Washington Post. Retrieved February 6, 2017 from https://www.washingtonpost.com/news/speaking-of-science/wp/2017/02/06/dear-science-why-do-we-love-our-pets/?utm_term=.8bb74c826ed7.

Kawasaki, G., & Fitzpatrick, P. (2014). *The art of social media: Power tips for power users*. London, UK: Penguin.

Keeling, L. J. (2005). Healthy and happy: Animal welfare as an integral part of sustainable agriculture. *AMBIO: A Journal of the Human Environment, 34*(4), 316–319.

Kitsak, M., Gallos, L. K., Havlin, S., Liljeros, F., Muchnik, L., Stanley, H. E., et al. (2010). Identification of influential spreaders in complex networks. *Nature Physics, 6*(11), 888–893.

Kwak, H., Lee, C., Park, H., & Moon, S. (2010, April). What is twitter, a social network or a news media? In *Proceedings of the 19th international conference on worldwide web* (pp. 591–600). New York: ACM.

Liu, W., Sidhu, A., Beacom, A. M., & Valente, T. W. (2017). Social network theory. In *The international encyclopedia of media effects*. Hoboken, NJ: John Wiley & Sons, Inc.

Long, R. (2016). Why are cosmetics industries using animals to test their products? Odyssey. Retrieved August 30, 2016 from: https://www.theodysseyonline.com/why-are-cosmetics-industries-using-animals-to-test-their-products.

Łopaciuk, A., & Łoboda, M. (2013). Global beauty industry trends in the 21st century. In *Management, knowledge and learning international conference*. Zadar, Croatia.

Lovejoy, K., & Saxton, G. D. (2012). Information, community, and action: How nonprofit organizations use social media. *Journal of Computer-Mediated Communication, 17*(3), 337–353.

Lu, J., Bayne, K., & Wang, J. (2013). Current status of animal welfare and animal rights in China. *ATLA, 41*, 351–357.

Luckerson, V. (2016). Here's proof that instagram was one of the smartest acquisitions ever. Time. Retrieved April 19, 2016 from http://time.com/4299297/instagram-facebook-revenue/.

McCarthy, A. (2016). Millennials dominate us beauty market. eMarketer. Retrieved December 14, 2016 from https://www.emarketer.com/Article/Millennials-Dominate-US-Beauty-Market/1014857.

Morell, Vi. (2014). Causes of the furred and feathered rule the internet. Retrieved March 14, 2014 from http://news.nationalgeographic.com/news/2014/03/140314-social-media-animal-rights-groups-animal-testing-animal-cognition-world/.

Nars. (2017). @narsissist on Instagram. Retrieved July 28, 2017 from https://www.instagram.com/p/BV2Qf8MjIi3/?hl=en.

Nonprofit Technology Network. (2011). Nonprofit social network benchmark report. Retrieved from http://www.NonprofitSocialNetworkSurvey.com.

Ormandy, E. H., & Schuppli, C. A. (2014). Public attitudes toward animal research: A review. *Animals, 4*(3), 391–408.

People for the Ethical Treatment of Animals. (2012). Chinese scientists learn non-animal testing, Thanks to PETA. Retrieved November 7, 2012 from https://www.peta.org/blog/chinese-scientists-learn-non-animal-testing-thanks-peta/.

People for the Ethical Treatment of Animals. (2014). New study shows grwoing opposition to animal tests. Retrieved February 16, 2014 from https://www.peta.org/media/news-releases/new-study-shows-growing-opposition-animal-tests/.

People for the Ethical Treatment of Animals. (2016). Update: China to approve first non-animal cosmetics test. Retrieved November 2, 2016 from https://www.peta.org/blog/china-approve-first-non-animal-cosmetics-test/.

People for the Ethical Treatment of Animals. (2017). Home. Retrieved from https://www.peta.org/.

Ramli, N. S. (2016). Green marketing: A new prospect in the cosmetics industry. In A. J. Vasile & D. Nicolò (Eds.), *Sustainable entrepreneurship and investments in the green economy* (pp. 200–230). Hershey, PA: IGI Global.

Russell, W., & Burch, R. (1959). *The principles of humane experimental technique*. London UK: Methuen.

Salemme, N. (2017). From Nars to Covergirl: The shocking truth behind your favourite makeup brands. Retrieved September 7, 2017 from http://www.news.com.au/lifestyle/beauty/face-body/from-nars-to-covergirl-the-shocking-truth-behind-your-favourite-makeup-brands/news-story/fc7b5d8076d1c8daea20b00737b2517d.

Saltzman, S. (2017). NARS takes to Instagram to address the issue of animal testing in China. Fashionista. Retrieved June 27, 2017 from https://fashionista.com/2017/06/nars-available-china-animal-testing-cosmetics.

Scholtz, B., Burger, C., & Zita, M. (2016). A social media environmental awareness campaign to promote sustainable practices in educational environments. In *Advances and new trends in environmental and energy informatics* (pp. 355–369). New York City, NY: Springer International Publishing.

Scott, J., & Carrington, P. J. (2011). *The SAGE handbook of social network analysis*. Thousand Oaks, CA: SAGE publications.

Sheble, L., Brennan, K., & Wildemuth, B. (2016). Social network analysis. In B. Wildemuth (Ed.), *Applications of social research methods to questions in information and library science* (2nd ed., pp. 339–350). Santa Barbara, CA: ABC-CLIO.

Sheldon, P., & Bryant, K. (2016). Instagram: Motives for its use and relationship to narcissism and contextual age. *Computers in Human Behavior, 58*, 89–97.

Shirky, C. (2011). The political power of social media: Technology, the public sphere, and political change. *Foreign affairs, 90*(1), 28–41.

Son, C. G. (2016). Dynamic of low-income households using social network analysis. *Journal of Korea Institute for Health and Social Affairs, 9*, 251–260.

Srivastava, J., Ahmad, M. A., Pathak, N., & Hsu, D. K. W. (2008). Data mining based social network analysis from online behavior. In *Tutorial at the 8th SIAM international conference on data mining* (SDM'08).

Statista. (2017). Leading beauty brands ranked by number of instagram followers as of August 2017. Retrieved November 4, 2017, from https://www.statista.com/statistics/536991/leading-beauty-brands-instagram-followers/.

Statista. (2017). Number of monthly active instagram users from January 2013 to September 2017. Retrieved November 4, 2017, from https://www.statista.com/statistics/253577/number-of-monthly-active-instagram-users/.

Stelzner, M. A. (2013). *Social media marketing industry report: How marketers are using social media to grow their businesses*. SocialMediaExaminer.com.

The Humane Society of the United States. (2017). Cosmetic testing on animals. Retrieved from http://www.humanesociety.org/assets/pdfs/animals_laboratories/cosmetic_product_testing/ARI-timeline-flier.pdf.

Thelwall, M., Buckley, K., & Paltoglou, G. (2011). Sentiment in Twitter events. *Journal of the Association for Information Science and Technology, 62*(2), 406–418.

United States Congress. (2017). H.R.2790—Humane Cosmetics Act. Retrieved from https://www.congress.gov/bill/115th-congress/house-bill/2790.

United States Department of Agriculture. (2017). Animal Welfare Act. Retrieved August 30, 2017, from https://www.nal.usda.gov/awic/animal-welfare-act.

Wasserman, S., & Faust, K. (1994). Social network analysis: Methods and applications (Vol. 8). Cambridge: Cambridge university press.

Wellman, B. (2001). Computer networks as social networks. *Science, 293*(5537), 2031–2034.

Wikipedia. (2017, April 10). United Express Flight 3411 incident. Retrieved from Wilkinson, J. (2014, February 19). Is social media impacting animal testing? Reset. Retrieved from https://en.reset.org/blog/social-media-impacting-animal-testing-02192014.

Wright, D. K., & Hinson, M. D. (2011). A three-year longitudinal analysis of social and emerging media use in public relations practice. *Public Relations Journal, 5*(3), 1–32.

Yan, S. (2017). In China, big cosmetics firms are selling products tested on animals. Retrieved April 19, 2017 from https://www.cnbc.com/2017/04/19/in-china-big-cosmetics-firms-are-selling-products-tested-on-animals.html.

Zuber, M. (2014). A survey of data mining techniques for social network analysis. *International Journal of Research in Computer Engineering & Electronics, 3*(6), 1–8.

Chapter 7
Cashmere Industry: Value Chains and Sustainability

Sheikh I. Ishrat, Nigel P. Grigg, Nihal Jayamaha and Venkateswarlu Pulakanam

"I do not believe in God. I believe in Cashmere."
(Fran Lebowitz)

Abstract As one of the luxurious fibres, cashmere is a fine natural fibre extracted from the underbelly of a unique goat primarily found in Asia. Cashmere fibre is used to make a variety of fashion and luxury woollen products such as scarves, shawls, pullovers and cardigans. Similar to fine wine or cheeses, it can be argued that the inherent value of cashmere to the consumer relates to both the properties of the product itself (softness, fineness), and the traditions and practices associated with its production. India is the traditional hub of cashmere products for centuries, and the handmade Indian cashmere products are unmatched in artisan expertise, range, design, appearance and quality. In the last few decades, due to technological advancements in manufacturing practices, the automation of cashmere manufacturing is fast replacing the centuries-old traditional practices. As a result of these changes, the industry in India is currently facing sustainability issucs. This research aims to determine how, and to what extent, automation is impacting traditional cashmere manufacturing processes and how these practices can be integrated for sustainability. To achieve this, we intend to explore cashmere luxury value chains and study the impact of automation on the value generation stages of these chains. This study will be carried out in the context of traditional craft industry making

S. I. Ishrat (✉)
Ara Institute of Canterbury, Christchurch, New Zealand
e-mail: Imran.ishrat@ara.ac.nz

N. P. Grigg · N. Jayamaha
Massey University, Palmerston North, New Zealand
e-mail: N.Grigg@massey.ac.nz

N. Jayamaha
e-mail: N.P.Jayamaha@massey.ac.nz

V. Pulakanam
University of Canterbury, Christchurch, New Zealand
e-mail: venkat.pulakanam@canterbury.ac.nz

© Springer Nature Singapore Pte Ltd. 2018

C. K. Y. Lo and J. Ha-Brookshire (eds.), *Sustainability in Luxury Fashion Business*, Springer Series in Fashion Business, https://doi.org/10.1007/978-981-10-8878-0_7

transition to the global, consumer age. Finally, based on the research aim, propositions are presented to address and integrate these aspects for future research. Through this work, a contribution to the body of knowledge surrounding cashmere industry is expected.

Keywords *Pashmina* · Cashmere · Value chains · Sustainability

Introduction

Cashmere—also known as *Pashmina* or *Pashm*—is an extremely fine natural fibre extracted from the underbelly of a unique goat primarily found in Asia. Contrary to general belief, cashmere is hair and not wool. Cashmere hair or fibre is used to make a variety of woollen products such as scarves, shawls, pullovers and cardigans. Similar to fine wine or cheeses, it can be argued that the inherent value of cashmere to the consumer relates in part to the properties of the product itself (in this case softness and, fineness), and in part to the traditions and practices associated with its production, which help develop those qualities.

Over a period of time, cashmere products have reached most other parts of the world and are well known for their fineness, warmth, softness and elegance (Waldron, Brown, & Komarek, 2014). Early instances of cashmere products include *Kashmiri* shawls reaching Britain and parts of Europe in the mid-eighteenth century via officials of the East India Company (Zutshi, 2009), and the Emperor Napoleon Bonaparte presenting an 'exquisite' *Kashmiri* shawl to Empress Joséphine de Beauharnais, who subsequently became passionate about similar articles (Ashraf, Ashraf, & Hafiz, 2016). At present, some European countries not only produce cashmere, but also import cashmere and its products in significant quantities from overseas markets which have resulted in a multi-billion dollar global cashmere trade (Waldron et al., 2014; Berger, Buuveibaatar & Mishra, 2013). Countries such as China and Mongolia, along with few other Asian countries such as Afghanistan, Kazakhstan, Iran and India, are foremost in producing different grades of cashmere fibres and products. Britain, Australia and New Zealand also produce cashmere fibre, however, in relatively small quantities (Shakyawar, Raja, Kumar, Pareek, & Wani, 2013). Globally, the annual production of cashmere is negligible in the world textile market; however, almost all major fashion brands include some cashmere products in their product line as part of their luxury components.

In general, a supply chain can be considered as a set of activities involved in meeting customer's requirements through delivering goods and services (Chopra & Meindl, 2016). In a traditional supply chain; suppliers, manufacturers, distributors and retailers plan and coordinate their efforts for the conveyance of goods in the form of raw materials, work-in-process and final deliverables. In doing so, value is created along the chain, thus creating a value chain. In cashmere supply chains, value is created in all the supply chain stages (Sect. 2), and in each stage, there is a network of organizations which potentially can generate value towards the end product.

A generic cashmere supply chain is classified into four broad stages: fibre procurement, fibre processing, fibre transformation and product labelling. As opposed to the conventional classification of supply chain activities mentioned earlier, it is more pertinent to classify and view the cashmere supply chain through these stages due to their significance in capturing and establishing luxury value to the end consumer. Recycling cashmere may provide a new dimension to the cashmere supply chain. However, recycling cashmere products is a relatively evolving concept, and in the literature little attention is drawn on this vital dimension of the chain. This perhaps is due to the fact that it is unlikely to recycle cashmere fibre to produce handmade luxury products without affecting its quality (Ashraf et al., 2016). However, due to the emergence and significance of corporate social responsibility in the textile industry (Wiengarten, Lo & Lam, 2017), few luxury brands do provide product take-back mechanisms to promote cashmere recycling towards blended products made through automated processes, but at present, such efforts are limited. In this regard, more attention is required by the researchers in future studies to explore recycling possibilities in cashmere operations.

In this study, the term 'supply chain' is used to denote the physical operations and distribution channels relating to the supply of cashmere products to end customers, whereas the term 'value chain' refers to the processes and stages that add luxury value to cashmere fibres as they pass through the supply chain's various operations. The human contribution (expertise, gender, tradition) is significant in generating value in cashmere industry, but is seldom considered or appreciated. To fully understand the phenomenon of cashmere, it is imperative for researchers to consider these aspects.

Cashmere: Value Generation Stages

Value is a broad and abstract concept and can be perceived in more than one way if viewed from the lenses of: technical, organizational and end-consumer value (Ramsey, 2005). Simplistically, value can be defined as the amount customers are willing to pay for a commodity that an organization provides (Porter, 1985). Within an organization, activities and relationships in and between various functional units generate value for customers by successfully managing various departments' activities (Porter, 1985). In regard to value chains, it can be considered as a coordinated set of activities between various inter- and intra-firm linkages to generate value towards the deliverable (Porter, 1990).

In the literature, value chains are primarily addressed from two aspects: a network- or chains-based aspect and a value generation aspect. Network- or chains-based approach considers strategic horizontal and vertical alliances of organizations to deliver products or services to the end-user (Trienekens, 2011). The value generation approach considers the customers' perspectives in creating value towards the product all along the chain (Wang & Wu, 2012). Overall, the value chain framework encompasses organizational networks and linkages in regard to sharing of resources to generate value for the local, regional or global clientele (Gereffi, 1994).

In the cashmere industry, the value of the final deliverable is likely to vary for different customers and markets. For instance, a genuine Kashmiri shawl may fetch different amounts in domestic and international markets depending on the monetized value a customer is willing to pay for such an article. Value, however, has another dimension, i.e. perceived quality, or aesthetics—one of Garvin's originally specified dimensions of product quality (Garvin, 1987). It is not always a direct indicator of the monetary value. In the literature, value chain studies have been conducted in regard to such products as cheese (Schmitt, Keech, Maye, Barjolle, & Kirwan, 2016), sugar (Higgins, Thorburn, Archer, & Jakku, 2007), cocoa (Haynes, Cubbage, Mercer, & Sills, 2012) and silk (Patichol, Wongsurawat, & Johri, 2014). We now discuss the four major cashmere value chain stages, highlighting practices and trends that are of relevance to the key themes of quality, value and sustainability.

Stage 1: Procurement

Geographical Area

Cashmere is the most expensive nature fibre (Ryder, 1984). The finest quality of cashmere fibre is produced in India with an annual production of about 40 tonnes harvested from nearly 200,000 cashmere goats which are endemic to the northern parts of the country (Ganai, Misra, & Sheikh, 2011). Unlike China, Mongolia and Afghanistan, India contributes less than one per cent of global cashmere production (Shakyawar et al., 2013). Importantly, however, the fibre quality and the products being made in *Kashmir*, India makes it particularly unique and much sought after in the domestic, regional and international markets. In India, cashmere is also known as *Pashmina* or *Pashm* or 'soft gold' (Bumla et al., 2012; Yaqoob, Sofi, Wani, Sheikh, & Bumla, 2012). Jammu and Kashmir (J&K) is the northernmost Indian state which has three major regions: Jammu region, Kashmir Valley and Ladakh. *Pashmina* is primarily procured from eastern Ladakh, a region close to the Tibetan Plateau. *Pashmina* is also procured from certain areas of Himachal Pradesh and Uttarakhand, the other two states in north India (Wani, Wani, & Yusuf, 2009).

Types of Goat Breeds

In India, raw cashmere fibre is primarily collected from two cashmere goat breeds: *Changthangi* (Ammayappan, Shakyawar, Krofa, Pareek, & Basu, 2011) and *Chegu* (Wani et al., 2009): former is also known as *Changra* (Namgail, Van Wieren, & Prins, 2010). However, apart from these breeds, *Pashmina* is also procured from *shapo* (*Ovis orientalis*) and Himalayan ibex (*Capra ibex*) based on the market requirements. The raw fibre is procured when the herbivores naturally shed their undercoat during the moulting season between March and May (Yaqoob et al., 2012). Among these goat breeds, *Changthangi* (inhabited to Ladakh) contributes more than three-quarters of the total *Pashmina* being procured in India. Furthermore, the average fibre diameter of *Changthangi* breed is 10–14 μ, which is significantly less than the prescribed classification of 19 μ for fibre to be considered as *Pashmina* which makes *Changthangi* procured fibre as one of the finest raw cashmeres in the world (Wani et al., 2009; Raja, Shakyawar, Pareek, Temani, & Sofi, 2013; Shakyawar, Raja, Wani, Kadam & Pareek, 2015). Also, the fibre length of *Changthangi* fibre is between 55 and 60 mm, making it one of the most sought-after fibres in the industry (Shakyawar et al., 2013). The exquisiteness of *Changthangi Pashmina* makes it an ideal process to be examined for the purposes of the present study.

Fibre Quality

Among other characteristics such as fibre strength, length and colour, the impact of weathering is also considered to ensure the quality of the procured fibre. Weathering of cashmere fibre is the result of Cashmere goat grazing in natural pastures for long durations, and it may lead to degradation of fibre softness and curvature which is significant in determining the fibre quality (McGregor, 2016). Cashmere goat is a relatively scarce resource from which cashmere fibre can be procured. Inconsistent or non-standardized procurement practices may lead to inferior fibre quality. Moreover, quality issues upstream in the chain may further affect the cashmere procurement cycle leading to sustainability issues. Determining fibre quality is a challenging and tedious process, but fibre quality is essential to the quality of the end product. In the fashion industry, established brands are conscious of the significance of the fibre quality and the value it generates towards the final product.

Stage 2: Processing

Combing

Raw cashmere contains naturally occurring contaminants such as dirt, sand, skin flakes and wax which can adhere to the procured fibre (Shakyawar et al., 2013). In combing, these impurities are eliminated. Traditionally, combing is carried out manually using a wooden comb (Shakyawar et al., 2013).

Dehairing

Cashmere fibre is harvested using combing from the goats once a year. On an average, the goat produces no more than 250 g of raw fibre annually (Namgail et al., 2010). The raw fibre consists of two components: the outer guard hair and the inner coat of fine fibre with a proportion of approximately 60–40%, respectively. The harvested cashmere fibre is then subject to a cleaning process called dehairing, in which the fine cashmere fibre is separated from the outer coarse hair. As cited in Shakyawar et al. (2015), according to Franck (2001), the goat down (inner coat) is six times finer than the human hair. For quality products, the guard hair must completely be separated from the fine inner coat before further processing. In case of more than 5% of guard hair in the fibre, the fibre quality is severely affected and degrades the quality of the final deliverable (Shakyawar et al., 2013).

Traditionally, dehairing is carried out through a time-consuming and labour-intensive manual process. Nowadays, dehairing is also subject to mechanical processes to speed up the complex tasks and is replacing the tedious manual effort. However, with the mechanized process, the dehaired fibre is observed to be of inferior quality in regard to fibre strength and tenacity (Shakyawar et al., 2013).

Animal Rights

Animal rights is a fundamental element in operations involving fauna in cashmere industry. However, unfortunately, limited literature is available addressing this important upstream cashmere supply chain aspect. Although few attempts are made to address *Pashmina* kid goats towards their husbandry practices, feed (Singh, Verma, Dass, & Mehra, 1999; Zhang et al., 2009) and livestock management (Baba, Wani & Zargar, 2011; see also Sect. 3.7), surprisingly not much attention is given on other animal rights related aspects for cashmere goats.

Stage 3: Transformation

Spinning

Once fibre processing is complete, the next step is to spin the fibre to manufacture yarn. The manufacturing of cashmere products, especially shawls, is carried out exclusively in the Kashmir Valley—more specifically in Srinagar City. Over the years, cashmere yarn manufacturing techniques have evolved significantly and are carried out through various ways to cater the requirements of different market segments. Cashmere yarn is made from pure cashmere fibre and through a blend of other natural and synthetic fibres to suit the product requirements (Shakyawar et al., 2013).

Traditional Approaches

In India, handmade cashmere products have been made by the local manufacturers and artisans to cater the needs of domestic as well as international markets since the eighteenth century (Ammayappan et al., 2011). Traditionally, cashmere yarn has been spun by hand on a spinning wheel called '*Charkha*' or on a more recently invented pedal-operated *Charkha* (Wani et al., 2013). Once spinning is over, weaving is carried out on the handlooms. In J&K alone, there are 478 handloom cooperative societies registered with a membership of more than 15,000 and 20,000 weavers in the organized sector and unorganized sector, respectively (Ahmad & Nengroo, 2013). Furthermore, not only has the handloom industry been one of the most significant contributors for employment and revenue generation in the state, but it also has no adverse impact on the environment (Ahmad & Nengroo, 2013). The handloom woven products are made by skilled artisans and generate more value to the product (Shakyawar et al., 2013).

Interestingly, at present, not many cashmere products are handmade except for the ones being made in Kashmir, India, and in some other parts of the world. However, the sophistication, aesthetics, expertise and artistic value traditionally made *Kashmiri Pashmina* products enjoy globally is unparalleled.

Recent Developments

In the last few decades, a significant increase in the demand for cashmere products has been observed globally. At the same time, technological advancements in manufacturing practices led to automated cashmere manufacturing processes since the traditional way of processing cashmere is *time-consuming* (Raja et al., 2013), *inconsistent* and *expensive* (Nibikora & Wang, 2010). As a result, the automation of cashmere manufacturing is fast replacing the centuries-old traditional practices resulting in a paradigm shift in various stages of cashmere processing. Besides cashmere, there is a similar emerging trend observed in other natural products such as milk where synthetic equivalents are being produced for global markets which will have an impact on cows and other milk-producing herbivores and may affect the industry significantly (Mudgil & Barak, 2013). Synthetic meat patties are being tested in laboratories before they make a commercially viable option for global consumption in food industry (Heid, 2016). Shifting trends from consuming original products to polymer-based ones may open new markets and capture larger customer base, but their impact in regard to sustainability aspects is yet to be identified and leaves scope for research.

Blending

There is a trend to automate cashmere yarn manufacturing process since it is significantly less expensive in time and cost as compared to the traditional handmade processes (Bumla et al., 2012). Cashmere fibre is very delicate and cannot be processed on the machines in its natural form (Shakyawar et al., 2015). For this reason, cashmere fibre is spun on the machines after blending with other fibres (natural and synthetic) to make it robust enough to sustain the stress of the machining process (Shakyawar et al., 2015). In spite of blending, processing cashmere fibre via the automated processes is significantly challenging. Inevitably, the quality of cashmere products produced from the automated process is different from the quality of products made completely by hand (i.e. without blending). This has resulted in experimentation with various blends of fibre, as discussed next.

Blending with Natural Fibres

Shakyawar et al. (2015) study the impact of blended cashmere fibre with natural fibres such as wool, rabbit hair and silk. Cashmere fibre has the same chemical composition as sheep wool; however, its structural properties are considerably different with regard to fibre uniformity and evenness (Raja et al., 2013). Cashmere and wool not only have similar chemical properties but also have similar morphological structure which makes cashmere and wool a natural choice for blending. Moreover, it is observed that up to 30% blending can be achieved without significantly compromising on fibre quality (Shakyawar et al., 2013). As a result, a blended cashmere–wool ratio can be obtained in different proportions such as 80:20, 70:30 and, at times, at 50:50 which reduces the costs up to 40% depending on the ratio used in the blended fibre (Shakyawar et al., 2013). Cashmere–wool blend is considered by Wang, Chang, and McGregor (2006) in regard to fibre curvature on hairiness, whereas McGregor and Schlink (2014) study the same blend to study the impact of nutrition on feltability of the blended fibre. The focus of the work by Naebe and McGregor (2013) is on the comfort properties of the knitted fabric of the cashmere–wool blend ratios and fibre curvature. Image processing technique is used by Qian, Li, Cao, Yu and Shen (2010) to calculate cashmere–wool blend yarn ratio. The blend of cashmere with another natural fibre, namely silk, is considered by Nibikora and Wang (2010) to identify the optimum settings in regard to different yarn spinning techniques.

Blending with Synthetic Fibres

Few attempts have been made by researchers to study the blend of cashmere with synthetic fibres. A study with regard to cashmere–nylon blend is considered by Raja, Shakyawar, Pareek and Wani (2011) to make the yarn robust enough to be used in the automated cashmere spinning process. Efforts have been made to blend cashmere with polyvinyl alcohol (PVA) to make it suitable for machining process (Raja et al., 2013).

Blending Yarn Treatment

Once blending is over, the blended yarn is treated with a chemical process to eliminate the external fibre component without significantly affecting the properties of the natural fibre (Shakyawar et al., 2015). In the literature, an attempt is made by Raja et al. (2011) to study the impact of nylon dissolution on shawls made from the blended cashmere–nylon yarn whereas the work of Raja et al. (2013) considered the effects of eliminating PVA portion from the finished cashmere products.

Dyeing

The dyeing process is carried out manually using organic compounds to add further value to the product. For instance, saffron-based dyes, made from saffron (Crocus sativus linn) flower, are used as natural colour to dye *Pashmina* (Raja et al., 2012). Similarly, walnut (Juglans nigra) husk is used in dying *Pashmina* shawls and similar articles (Lal et al., 2011). Another natural dye extract is used which is obtained from a small spreading tree (Kigelia pinnata) to dye *Pashmina* fabrics (Sharma et al., 2013). However, natural dyes are expensive and moderate in fastness (Umbreen, Ali, Hussain & Nawaz, 2008). This aspect of natural dyes may make them less attractive for large-scale operations, and their usage without proper measures can vary in colour consistency and eventually compromise on the product quality. As a result, various chemical or synthetic dyes are also used for dyeing, primarily to lower the costs associated with this process (Sheikh, 2014).

However, in recent years, environmental awareness has attracted the attention of manufacturers (Geelani, Ara, Mir, Bhat, & Mishra, 2016). As a result, to some extent, natural dyes are preferred over the synthetic ones (Patchaiyappan & Yogamoorthi, 2014). Moreover, the persistent use of synthetic dyes results in carcinogenic effects (Kulkarni, Bodake, & Pathode, 2011), toxic discharge (Padmavathy et al, 2003), water contamination (Prado, Torres, Faria, & Dias, 2004) and may have detrimental effects on the ecology and human health.

Weaving

Post dyeing, yarn is weaved as warp. This tedious process is carried out in many stages and takes many days to accomplish. During weaving, one of the most challenging and probably frustrating tasks is to make sure that no sooner does the thread break on the loom (which happens frequently), the weaver picks it and replaces it with a new one. In this process, in general, one-tenth of waste is considered within the acceptable range (Yaqoob et al., 2012).

Stage 4: Labelling

Products

Among all the cashmere products, *Kashmiri* shawls are the most sought-after commodity. Handmade genuine cashmere shawls made in India are recognized globally and are considered as the most profitable proposition for the manufacturers and artisans alike. There are different types of handmade shawls made through various embroidery styles such as *Kani, Sozani, Kanizamar* (Shakyawar et al., 2013). These articles, if taken care of well, last more than two decades (Bumla et al., 2012) and later used as antiques and souvenirs for future generations. Apart from shawls, a significant variety of other cashmere products are made including scarves, stoles, pullovers and blankets.

Standards

In the fashion industry, cashmere is a well-known product and luxury brands include cashmere products in their product lines to cater to the requirements of their clientele by designing and developing value chains. In cashmere value chains, specialty retailers (specific product sellers) and mass merchant retailers (multi product sellers) play significant role in establishing the value of the end product to the consumer. Labelling is critical in fashion industry in establishing the value for the customer. However, unscrupulous manufacturers and distributors do not conform to the standards prescribed by the Cashmere & Camel Hair Manufacturers Institute (CCMI)—an international organization representing cashmere and other natural fibres. Appropriate labelling on cashmere products is a significant issue in the industry, and there are few standards available such as ISO 1005-A02 and BS 1006-B02 (colour fastness) to streamline some aspects of cashmere processes globally (Raja & Thilagavathi, 2008).

In Table 7.1, a summary of the cashmere supply chain phases and activities involved in each phase in regard to traditional and modern practices is presented.

Cashmere: Sustainability Issues

Currently, the cashmere industry is operating in a challenging environment. In this section, significant issues that face the present-day cashmere industry are presented. Similar to value, sustainability is a broad and complex concept rather than a problem-solving approach. As a result, first of all, the term 'sustainability' needs to be understood, especially in the context of the study. In the literature, sustainability is defined in many ways. One of the most comprehensive and widely accepted views on sustainability is defined in the Brundtland Report (1987) on World

Table 7.1 Comparison of traditional and modern cashmere-related practices

Value chain			Practice	
Stage	Activity	Subactivity	Traditional	Modern
Procurement	Fibre collection		Manual	Manual
	Dehairing	Combing	Wooden comb Yak horn	Steel comb Shearing automated
Processing	Cleaning	Dirt sand Skin flakes	Manual	Automated
	Sorting	Colour Length Curvature	Manual	Automated
Transformation	Blending	Natural	Manual	Manual
		Synthetic		
	Spinning	Natural	Manual	Automated
		Blended		
	Weaving		Manual	Automated
	Dyeing		Manual	Manual
Labelling	Product	Shawl	Manual	Automated
		Stole		
		Scarf		
		Pullover		
		Blanket		
Recycling		Shawl Scarf		Automated

Commission on Environment and Development as 'meeting the needs of the present without compromising the ability of future generations to meet their own needs'. However, over the years, the concept of sustainability has evolved to capture multifaceted real-world phenomena through various drivers (ecological, social and economic) at different physical (infrastructure) and conceptual (cultural) levels.

In the context of the present study, the significance and impact of the sustainability aspects are important to understand. These aspects can be studied in two dimensions: temporal (cashmere practices before and after automation) and geographical (specifically in relation to India). Integrating sustainability into various cashmere industry aspects includes the consideration of livestock management and husbandry practices, herders, procurement techniques, fibre processing, yarn manufacturing, product design, distribution channels and standards. The entire gamut of cashmere-related activities establishes a considerable scope to study sustainable cashmere processes.

Rise of Unscrupulous Manufacturing Practices

In the cashmere industry, the end-users play significant role in determining the product quality. In the last few decades, this driver has forced the cashmere industry to adopt unconventional practices to cater to the requirements of the broad customer base for cashmere products. Due to this, in many instances, unscrupulous manufacturing practices have evolved which have compromised the product quality (Shakyawar et al., 2015).

Lack of Industry Standards and Controls

The lack of industry standards, coupled with non-compliance, escalates the cost of quality at various cashmere processing stages, eventually affecting the product value (Ashraf et al., 2016). Surprisingly, not many industry standards are in place to streamline various cashmere processes resulting in unethical practices which have significantly affected the industry (Shakyawar et al., 2015). In many cases, cashmere articles are not distributed or sold with proper labelling of the product, which misleads the consumer. For instance, genuine Kashmiri shawls have no trademark or patent to distinguish with their imitations (Ashraf et al., 2016). This encourages manufacturers, mostly based outside the Kashmir region, to produce and sell fake products with inappropriate labelling to capture the domestic and international markets (Ashraf et al., 2016).

Impact of Automation on Social Dynamics

Based on the recent 2011 census in India, 12.5 million inhabitants live in the Indian state of J&K, of which more than two-thirds live in the rural areas. It is worth mentioning that the vast majority of the rural population in J&K relies on professions requiring skilled and manual work, mainly using natural resources for sustenance (Ramasubramanian, 2004). In J&K, one of the most common and traditional industries is *Pashmina* which has changed considerably over the years due to the emergence of the state-of-the-art cashmere production practices. For instance, the increase in the productivity through automated processes has impacted the production of handmade Kashmiri shawls. In comparison with pre-automation era, the production of such articles has significantly declined (50,000 are produced annually) and has affected the livelihoods of more than 300,000 artisans and workers associated directly or indirectly with this industry (Sheikh, 2014). The manual process to produce cashmere articles is a time-consuming and tedious process. Unfortunately, the hard work of artisans is not reflected in their wages leading to dissatisfaction among the community. The social dimension of sustainability is not addressed as exclusively and comprehensively as the environmental and the economic dimensions (Cuthill, 2009). With the shift in the trends in cashmere manufacturing as mentioned earlier, upstream the chain, many herders have shifted their base to urban centres for reasons such as low-profit margins abandoning traditional pastoral practices which may create ecological imbalance in the region.

Impact of Automation on Environment

To a large extent, handicraft industries promote cleaner production. Handlooms consume significantly low energy than power looms and other automated processes (Ahmad & Nengroo, 2013). The shift towards automated cashmere processes results in industrial waste and emissions which is not addressed by manufacturers due to the lack of stringent policies and their implementation. For instance, in regard to blended fibre, almost all the outputs go through an extensive chemical dissolution process to eliminate the foreign element from the cashmere fibre (Raja et al., 2011, 2013). This chemical treatment is important to make the blended fibre 'look' as identical as possible to the natural cashmere fibre.

Market Segmentation and Financial Aspects

Based on the different ways of manufacturing a cashmere article, there are different markets to cater to the customers' requirements. For instance, a pure cashmere fibre shawl, made completely by hand, using traditional practices, has a niche market in India. Such an article is expensive and is more likely to establish a wider customer base in the international markets, including famous fashion brands and haute couture boutiques all over the world. On the contrary, a blended cashmere shawl made on automated processes is relatively inexpensive, in regard to time and cost, and would sell in the local market easily. In India, such a market is developing at an enormous pace and generates a considerable amount of revenue to manufacturers at a reasonably lower cost. There are different cashmere supply chains in action which lead to complexity in regard to manufacturing practices, conformance to standards and labelling issues in the chain. These chains use traditional or modern or a combination of both practices for domestic and international clientele. In cashmere industry, socially responsible products are exclusive to a niche market. Capturing this aspect, a study is considered by Ha-Brookshire and Norum (2011) to gauge the amount customers are willing to pay for cotton apparels made through sustainable practices. In regard to cashmere products, it would be interesting to see the degree to which consumers see value for such products in wider market segments.

Poor Communication Channels and Infrastructure

Due to dominant political factors prevailing in the area during the last few decades, there is limited infrastructural and industrial development in the state. As a result, there are a small number of viable options available for the people associated with cashmere industry to consider, but to primarily depend on the traditional ways to earn their livelihoods. For instance, telecommunication services such as Internet and mobile connectivity in the Kashmir region are government-controlled and are subject to scrutiny; leaving little room for the supply chain operators to rely on modern age communication channels to overcome the obstacles presented by traditional business practices.

Over Breeding of Cashmere Goats

Increasing demand for cashmere products is providing economic benefits to all the stakeholders in the chain. Due to the scarcity of raw material (cashmere fibre) and its commercial importance, there is a trend observed to increase the population of cashmere goats by the local populace for economic benefits in order to respond to increased fibre demand (Namgail et al., 2010). The impact of the shift is observed in the decline of number of sheep in the region (Baba et al., 2011). More demand for cashmere goats is leading to different husbandry practices, i.e. breed more goats (which is more profitable) than sheep. The change in the composition of livestock is changing the landscape of Kashmir region as there is now more need for plants and shrubs for the goats to graze than pasturelands which the sheep prefer. This affects the grazing pasturelands (Lin et al., 2012) and changes in the physical landscape of the region may eventually displace the wild herbivores such as Yak and *hangul* farther from their natural habitat (Namgail et al., 2010; Wehrden, Wesche, Chuluunkhuyag, & Fust, 2014), which as such is not sustainable for the ecosystem.

A comparative study conducted in India, Mongolia and China by Berger et al. (2013) states that the decline of large herbivores such as yak, saiga and chiru in the region is a result of increasing global cashmere markets since the fibre from these large herbivores is not as attractive as the ones procured from cashmere goats.

Proposed Research Scope

The global demand for cashmere products using pure and blended fibres is growing steadily (Ansari-Renani, 2014). In the present study, two cashmere value chains were compared: (a) value chain stages involving pure (*traditional chains*: Table 7.2) and (b) blended cashmere fibre (*modern chains*: Table 7.3). The traditional chains correspond to the practices adopted in each cashmere processing stage

Table 7.2 Overview of traditional cashmere value chain processes

Pre-automation era: traditional chains				
Procurement	Processing	Transformation		Labelling
Cashmere Goat		Artisans		Products
	Combing (wooden combs)	Fibre spinning (pure)	Dyeing	Shawl
	Dehairing	*Charkha* yarn (natural)	Weaving	Stole
			Finishing	Scarf
Farmers				Localized standards

Table 7.3 Overview of modern cashmere value chain processes

Post-automation era: modern chains				
Procurement	Processing	Transformation		Labelling
Cashmere goat		Artisans		Products
	Combing (automated)	Fibre spinning (blended)	Dyeing	Shawl Stole Scarf Pullover Blankets
	Dehairing (automated)	Machines	Weaving (automated)	
		Yarn (treated)	Finishing	
Cooperative or traders				Globalized standards

prior to the advent of mechanized processes where all the processes were carried out using centuries-old manual practices.

In an attempt to modernize tradition, in the modern chains, mechanized or automated processes have replaced many practices prevalent in the traditional chains. For instance, combing and dehairing are carried out using automated processes, and raw cashmere fibre is blended with synthetic fibres and spun on sophisticated machines. Furthermore, various new products such as pullovers and blankets are weaved on power looms and other mechanized processes. Also, electronic channels are considered to reach and capture wider markets.

In many instances, in these chains, processes do not conform to quality standards and threaten the sustainability of operations. We explore the possibility of capturing value generation stages in both cashmere chains. In the literature, there is no study conducted yet from the perspective of value chains to achieve sustainable cashmere operations.

Proposed Research Questions

We argue that an investigation of the cashmere problems through the lens of sustainable value chains and related aspects is critical. As discussed earlier, there are two dominant chains prevailing in the present-day cashmere industry. Currently, cashmere industry in India is not well organized, especially in regard to the *traditional chains*. It is undergoing a continuous change and significantly impacts the stakeholders involved with the industry, whereas the *modern chains* have evolved fast; however, the ripple effects of their practices are observed and reflected in the social, economic and environmental contexts. The focus of the research is on the traditional chains to study the impact of automation on the value generation stages of these chains.

Therefore, the central research aim that drives future inquiry is:

To determine how, and to what extent, automation is impacting traditional cashmere manufacturing processes in India and how these practices can be integrated for sustainability?

Based on this, the following propositions were developed:

- The automation in cashmere. Cashmere processing is significantly, both positively and negatively, impacting the social, economic and environmental aspects of the traditional cashmere value chains.
- Cashmere fibres are critical in creating value. Therefore, the current practice of using blended fibres may adversely affect quality, value or sustainability aspects.
- Globalization and related trends are, both positively and negatively, influencing indigenous cashmere production practices, which may impact quality, value and sustainability.
- Paradigm shift in the yarn manufacturing practices may have significant impact on the sustainability of the processes involved in traditional chains.
- In traditional chains, adaptation of the state-of-the-art techniques would help mitigate quality issues and lead to sustainable practices.

Because the cashmere industry's aim is to deliver luxury value, the findings will extend and build on previous research by two of the present authors that examined the use of Taguchi's Quality Philosophy (and related practices) in the context of a lean textile operation in Sri Lanka (Gamage, Jayamaha, & Grigg, 2016). As such, the present research intends to integrate the contemporary themes of quality, value and sustainability in cashmere apparel industry. These are viewed in the context of a developing economy and traditional craft industry making transition to the global, consumer age.

Conclusion

As observed in the literature, the cashmere industry in India is operating in a challenging environment and facing issues which impact the industry in many ways. In this study, we intend to capture processes and stages that add value to cashmere products as they pass through various supply chain operations. In this regard, the four major cashmere value chain stages are identified and will be addressed with the key themes of quality, value and sustainability. Based on the current trends, this is a crucial time to examine the cashmere industry and conduct an in-depth study on the impact that developments over the years have on the sustainability filters such as financial aspects, societal impact and environmental considerations on the region. As identified, in the present scenario, two cashmere chains are active: traditional chains and modern chains. We intend to explore these chains and study the impact of automation on the value generation stages from the lens of sustainability. The objective of proposed research is to explore the opportunity to identify sustainable value generation steps in cashmere value chains, and as such, the research will integrate the contemporary themes of quality, value and sustainability. In this novel study, theory surrounding cashmere operations will be developed in regard to various cashmere processing stages. Specifically, we expect to develop sustainable cashmere value chain, and through the research findings, contribute in establishing sustainable operations in cashmere industry. Furthermore, the stakeholders are expected to benefit from the findings of the study which is likely to impact cashmere industry practices in the region.

References

Ahmad, F., & Nengroo, A. H. (2013). An analysis of handloom sector of Jammu & Kashmir: A case study of district Budgam. *International Journal of Management and Business Studies, 3*(1), 106–109.

Ammayappan, L., Shakyawar, D. B., Krofa, D., Pareek, P. K., & Basu, G. (2011). Value addition of pashmina products: Present status and future perspectives—a review. *Agricultural Review, 32*(2), 91–101.

Ansari-Renani, H. R. (2014). A value chain and marketing of Iranian Cashmere. Media Peternakan: *Journal of Animal Science and Technology, 37*(1), 61–70.

Ashraf, S. I., Ashraf, S. N., & Hafiz, S. M. (2016). Obstacles faced by craftsmen and traders in pashmina sector: A study of J&K. *International Journal of Advanced Research, 4*(6), 1227–1239.

Baba, S. H., Wani, M. H., & Zargar, B. A. (2011). Dynamics and sustainability of livestock sector in Jammu & Kashmir. *Agricultural Economics Research Review, 24,* 119–132.

Berger, J., Buuveibbaatar, B., & Mishra, C. (2013). Globalization of the cashmere market and the decline of large mammals in Central Asia. *Conservation Biology, 27*(4), 679–689.

Bruntland Commission. (1987). *Our common future*. Oxford, United Kingdom: Oxford University Press.

Bumla, N. A., Wani, S. A., Shakyawar, D. B., Sofi, A. H., Yaqoob, I., & Sheikh, F. D. (2012). Comparative study on quality of shawls made from hand and machine spun pashmina yarns. *Indian Journal of Fibre & Textile Research, 37,* 224–230.

Chopra, S., & Meindl, P. (2016). *Supply chain management: Strategy, planning, and operation.* Boston, MA: Pearson.

Cuthill, M. (2009). Strengthening the social in sustainable development: Developing a conceptual framework for social sustainability in a rapid urban growth region in Australia. *Sustainable Development, 18*(6), 362–373.

Franck, R. R. (2001). *Silk, mohair, cashmere and other luxury fibres* (p. 142). Cambridge, MA: Woodhead. ISBN 1-85573-540-7.

Gamage, P., Jayamaha, N. P., & Grigg, N. P. (2016). Acceptance of Taguchi's quality philosophy and practice by Lean practitioners in apparel manufacturing. *Total Quality Management and Business Excellence, 28*(11–12), 1–17.

Ganai, T. A. S., Misra, S. S., & Sheikh, F. D. (2011). Characterization and evaluation of Pashmina producing Changthangi goat of Ladakh. *Indian Journal of Animal Sciences, 81*(6), 592–599.

Garvin, D. A. (1987). Competing in the eight dimensions of quality. *Harvard Business Review, 87* (6), 101–109.

Geelani, S. M., Ara, S., Mir, N. A., Bhat, S. J. A., & Mishra, P. K. (2016). Dyeing and fastness properties of Quercus robur with natural mordants on natural fibre. *Textiles and Clothing Sustainability.* https://doi.org/10.1186/s40689-016-0019-0.

Gereffi, G. (1994). *Introduction: Global commodity chains.* Westport, CT: Praeger.

Ha-Brookshire, J. E., & Norum, P. S. (2011). Willingness to pay for socially responsible products: Case of cotton apparel. *Journal of Consumer Marketing, 28*(5), 344–353.

Haynes, J., Cubbage, F., Mercer, E., & Sills, E. (2012). The search for value and meaning in the cocoa supply chain in Costa Rica. *Sustainability, 4*(7), 1466–1487.

Heid, M. (2016). You asked: Should I be nervous about lab-grown meat? Retrieved from http://time.com/4490128/artificial-meat-protein/. Accessed on February 23, 2017.

Higgins, A., Thorburn, P., Archer, A., & Jakku, E. (2007). Review: Opportunities for value chain research in sugar industries. *Agricultural Systems, 94,* 611–621.

Kulkarni, S. S., Bodake, U. M., & Pathode, G. R. (2011). Extraction of natural dye from Chili (Capsicum annum) for textile coloration. *Universal Journal of Environmental Research and Technology, 1,* 58–63.

Lal, C., Raja, A. S. M., Pareek, P. K., Shakyawar, D. B., Sharma, K. K., & Sharma, M. C. (2011). Juglans nigra: Chemical constitution and its application on Pashmina (cashmere) fabric as a dye. *Journal of Natural Product and Plant Resources, 1*(4), 13–19.

Lin, L., Dickhoefer, U., Müller, K., Wang, C., Glindemann, T., Hao, J … Susenbeth, A. (2012). Growth of sheep as affected by grazing system and grazing intensity in the steppe of Inner Mongolia, China. *Livestock Science, 144,* 140–147.

McGregor, B. A. (2016). Weathering, fibre strength and colour properties of processed white cashmere. *Journal of the Textile Institute, 107*(9), 1193–1202.

McGregor, B., & Schlink, A. (2014). Feltability of cashmere and other rare animal fibres and the effects of nutrition and blending with wool on cashmere feltability. *Journal of the Textile Institute, 105*(9), 927–937.

Mudgil, D., & Barak, S. (2013). Synthetic milk: A threat to Indian dairy industry. *Carpathian Journal of Food Science & Technology, 5*(1/2), 64–68.

Naebe, M., & McGregor, B. A. (2013). Comfort properties of superfine wool and wool/cashmere blend yarns and fabrics. *Journal of the Textile Institute, 104*(6), 634–640.

Namgail, T., Van Wieren, S. E., & Prins, H. H. (2010). Pashmina production and socio-economic changes in the Indian Changthang: Implications for natural resource management. *Natural Resources Forum, 34,* 222–230.

Nibikora, I., & Wang, J. (2010). Optimum selection of the opening roller and navel for rotor spun silk/cashmere blended yarn. *Fibres and Textiles in Eastern Europe, 82*(5), 35–38.

Padmavathy, S., Sandhya, S., Swaminathan, K., Subrahmanyan, Y. V., Chakrabarti, T., & Kaul, S. N. (2003). Aerobic decolorization of reactive Azo dyes in presence of various co substrates. *Chemical Biochemical Engineering, 17*(2), 147–151.

Patchaiyappan, A., & Yogamoorthi, A. (2014). Isolation, application and biochemical characterization of colour component from Tecoma stans: A new cost effective and eco-friendly source of natural dye. *International Journal of Natural Products Research., 4,* 9–11.

Patichol, P., Wongsurawat, W., & Johri, L. M. (2014). Modernizing tradition—Thai silk industry. *Strategic Direction, 30*(2), 31–33.

Porter, M. (1985). *Competitive advantage: Creating and sustaining superior performance*. New York, NY: Free Press.

Porter, M. (1990). *The competitive advantage of nations*. New York, NY: Free Press.

Prado, A. G. S., Torres, J. D., Faria, E. A., & Dias, S. C. L. (2004). Comparative adsorption studies of indigo carmine dye on chitin and chitosan. *Journal of Colloid and Interface Science, 1*(1), 43–47.

Qian, K., Li, H., Cao, H., Yu, K., & Shen, W. (2010). Measuring the blend ratio of wool/cashmere yarns based on image processing technology. *Fibres & Textiles in Eastern Europe, 18*(4), 35–38.

Raja, A. S. M., Pareek, P. K., Shakyawar, D. B., Wani, S. A., Nehvi, F. A., & Sofi, A. H. (2012). Extraction of natural dye from saffron flower waste and its application on pashmina fabric. *Advances in Applied Science Research, 3*(1), 156–161.

Raja, A. S. M., Shakyawar, D. B., Pareek, P. K., Temani, P., & Sofi, A. H. (2013). A novel chemical finishing process for cashmere/PVA-Blended Yarn-made cashmere fabric. *Journal of Natural Fibers, 10*(4), 381–389.

Raja, A. S. M., Shakyawar, D. B., Pareek, P. K., & Wani, S. A. (2011). Production and performance of pure cashmere shawl fabric using machine spun yarn by nylon dissolution process. *Indian Journal of Small Ruminants, 17,* 203–206.

Raja, A. S. M., & Thilagavathi, G. (2008). Dyes from the leaves of deciduous plants with a high tannin content for wool. *Coloration Technology, 124,* 285–289.

Ramasubramanian, R. (2004). *Can environmental security bring peace to Jammu and Kashmir.* New Delhi, India: Institute of Peace and Conflict Studies.

Ramsey, J. (2005). The real meaning of value in trading relationships. *International Journal of Operations & Production Management, 25*(6), 549–565.

Ryder, M. L. (1984). Prospects for cashmere production in Scotland. *Wool Record*, 37–43.

Schmitt, E., Keech, D., Maye, D., Barjolle, D., & Kirwan, J. (2016). Comparing the sustainability of local and global food chains: A case study of cheese products in Switzerland and the UK. *Sustainability, 8*(5), 419. https://doi.org/10.3390/su8050419.

Shakyawar, D. B., Raja, A. S. M., Kumar, A., Pareek, P. K., & Wani, S. A. (2013). Pashmina fibre—Production, characteristics and utilization. *Indian Journal of Fibre & Textile Research, 38,* 207–214.

Shakyawar, D. B., Raja, A. S. M., Wani, S. A., Kadam, V. V., & Pareek, P. K. (2015). Low-stress mechanical properties of pashmina shawls prepared from pure hand spun, machine spun and pashmina-wool blend yarn. *The Journal of the Textile Institute, 106*(3), 327–333.

Sharma, K. K., Pareek, P. K., Raja, A. S. M., Temani, P., Kumar, A., Shakyawar, D. B., et al. (2013). Extraction of natural dye from kigelia pinnata and its application on pashmina (cashmere) fabric. *Research Journal of Textile and Apparel, 17*(2), 28–32.

Sheikh, F. A. (2014). Exploring informal sector community innovations and knowledge appropriation: A study of Kashmiri pashmina shawls. *African Journal of Science, Technology, Innovation & Development, 6*(3), 203–212.

Singh, P., Verma, A. K., Dass, R. S., & Mehra, U. R. (1999). Performance of pashmina kid goats fed oak (Quercus semecarpifolia) leaves supplemented with a urea molasses mineral block. *Small Ruminant Research, 31,* 239–244.

Trienekens, J. H. (2011). Agricultural value chains in developing countries: A framework for analysis. *International Food and Agribusiness Management Review, 14*(2), 51–81.

Umbreen, S., Ali, S., Hussain, T., & Nawaz, R. (2008). Dyeing properties of natural dyes extracted from turmeric and their comparison with reactive dyeing. *Research Journal of Textile and Apparel, 12*(5), 1–11.

von Wehrden, H., Wesche, K., Chuluunkhuyag, O., & Fust, P. (2014). Correlation of trends in cashmere production and declines of large wild mammals: Response to Berger et al. 2013. *Conversation Biology, 29*(1), 286–289.

Waldron, S., Brown, C., & Komarek, A. M. (2014). The Chinese cashmere industry: A global value chain analysis. *Development Policy Review, 32*(5), 589–610.

Wang, H. W., & Wu, M. C. (2012). Business type, industry value chain, and R&D performance: Evidence from high-tech firms in an emerging market. *Technological Forecasting and Social Change, 79*(2), 326–340.

Wang, X., Chang, L., & McGregor, B. (2006). Hairiness of worsted wool and cashmere yarns and the impact of fiber curvature on hairiness. *Textile Research Journal, 76*(4), 281–287.

Wani, S. A., Sofi, A. H., Shakyawar, D. B., Yaqoob, I., Matto, F. A., & Malik, A. H. (2013). Fabrication of innovative charkha for pashmina spinning and its impact assessment. *The Journal of the Textile Institute, 104,* 1141–1144.

Wani, S. A., Wani, M. H., & Yusuf, S. (2009). Economics of Pashmina based trans-humane production system in cold arid region of Jammu and Kashmir. *Indian Journal of Agricultural Economics, 64*(2), 229–245.

Wiengarten, F., Lo, C. K. Y., & Lam, J. Y. K. (2017). How does sustainability leadership affect firm performance? The choices associated with appointing a chief officer of corporate social responsibility. *Journal of Business Ethics, 140*(3), 477–493.

Yaqoob, I., Sofi, A. H., Wani, S. A., Sheikh, F. D., & Bumla, N. A. (2012). Pashmina shawl–A traditional way of making in Kashmir. *Indian Journal of Traditional Knowledge, 11*(2), 329–333.

Zhang, W., Wang, R., Kleemann, D. O., Gao, M., Xu, J., & Jia, Z. (2009). Effects of dietary copper on growth performance, nutrient digestibility and fibre characteristics in cashmere goats during the cashmere slow-growing period. *Small Ruminant Research, 85,* 58–62.

Zutshi, C. (2009). "Designed for eternity": Kashmiri shawls, empire, and cultures of production and consumption in Mid-Victorian Britain. *Journal of British Studies, 48*(2), 420–440.

Chapter 8
Sustainability in the Fur Industry

Thomas C. C. Wong, Roger Ng and Lei Min Cai

Abstract A common understanding of "sustainability" is to manage the resources today so that future survival can be secured. Many people focus on the issue of environmental protection, and recycling. Yet, there are other perspectives of sustainability. Firstly, at the micro-level, will the demand of fur products as luxury products be sustainable? Are there any supporting reasons for the survival of the demand of fur products? Secondly, will the supply of fur products diminish? Are there evidences of the continuation of the supply? Will the man-made substitute of fur, namely faux fur, replace the genuine fur? Thirdly, at the macro-level, what kind of political impact will the fur industry create to help her own survival? Fourthly, how about the social and economic impact that the fur industry creates to ensure her existence? Is the economy of the fur-producing country heavily depending on the tax from fur industry? Is the employment also depending heavily on the fur industry? Consequently, is there a holistic model of sustainability in the fur industry? Will this model be able to explain how the existing resources affect the future survival? In this chapter, a qualitative holistic model of sustainability of the fur industry is proposed. The research method consists of both secondary data research and primary data that are acquired through interviewing delegates of different roles in the fur industry. The results were compared to check if they are coherent. Finally, the result was analyzed. The holistic model was extracted accordingly.

T. C. C. Wong (✉)
International Fur Federation, Block I, 6th floor, 36 Man Yue Street, Kowloon, Hong Kong
e-mail: thomas_c_c_wong@hotmail.com

R. Ng
Fashion & Textile Technology—Clothing, The Hong Kong Polytechnic University, Kowloon, Hong Kong
e-mail: roger.ng@polyu.edu.hk

L. M. Cai
Jinling Institute of Technology, Nanjing, China

© Springer Nature Singapore Pte Ltd. 2018

C. K. Y. Lo and J. Ha-Brookshire (eds.), *Sustainability in Luxury Fashion Business*, Springer Series in Fashion Business, https://doi.org/10.1007/978-981-10-8878-0_8

Introduction

The fur industry is a controversial heritage of the global economy. Fur has traditionally been a necessity for protecting against cold weather. However, today it is a symbol of luxury for most consumers. Few articles in the extensive literature on fashion industry sustainability have examined the fur industry (Islam & Mohammad, 2016; Lakshmanan, Jose, & Chakraborty, 2016; Routledge, 2014). Concerns about fashion industry sustainability have typically focused on environmental, carbon footprint during manufacturing, and wastage concerns. The trend in sustainability improvement is to investigate environmentally friendly manufacturing and chemical processes and promote minimal-waste designs or extend product life spans. Most evaluations have focused on the supply chain and environmental impacts by using life cycle analysis (Kozar & Hiller Connell, 2015; Roos, Zamani, Sandin, Peters, & Svanstro, 2016; Moller, 2008; Kozlowski, Searcy, Bardecki, 2015; Resta, Gaiardelli, Pinto, & Dotti, 2016). Therefore, the scope of the relevant literature has not been able to provide a holistic model of the sustainability of the fur industry.

Fur industry sustainability involves complex challenges. For example, some fashion designers prefer to use faux fur. The sustainability of the cotton supply is not a concern; by contrast, if certain furbearing animals are not protected properly, the fur supply might be disrupted. Retailers do not face activists blocking the entrances of their shops to protest the sale of wool jackets. Any sustainability model of the fur industry should therefore cover the entire supply chain together with political and economic considerations.

We have divided the content of this study into two areas: (1) sustainability of luxury fur products and (2) sustainability of the luxury fur industry. First, we consider the entire supply chain, starting with demand. Unlike other natural materials, such as cotton, flax, and wool, fur is controversial. What assurance is there that the demand for fur fashion products will continue? Additionally, raw material from furbearing animals must be available. In current practice, furbearing animals are either farmed or trapped. Only certain types of furbearing animals such as mink can be farmed, whereas other types such as skunk and otter can only be found in the wild. Will the supply of the raw materials remain available? More importantly, how are the animals being treated? Are the animal-raising processes ecologically sound? Are there any legal constraints or ethical considerations involved in the harvesting of furbearing animals? Moreover, commercial substitutes for fur (i.e., faux fur) have been developed to replace natural fur. Is this alternative more sustainable?

Next, we evaluate the relevant political and economic questions. Will governments abandon the fur industry and shut down its farms and factories? Will there be shortages of labor or fashion designers? What are the driving forces behind the continued supply of manpower in the industry?

In executing this study, we first conducted relevant literature reviews and interviews. Our primary sources included publications by the International Fur

Federation (IFF) and interviews with fur industry representatives. Our secondary sources included other publications and Web site pages, which provided an objective sense of the current situation in the fur industry so that the key factors affecting its sustainability could be identified. This enabled us to focus on interviewing fur industry representatives who could share their perceptions and strategies for shaping the future of the industry.

In the first section of this paper, we introduce the basic operations of the fur industry to help readers understand the logic of our results. We then explain our argument that fur products should be considered as luxuries instead of necessities. Subsequently, we analyze the foundation of fur product demand from two perspectives: fashion theory and cultural theory. Next, we introduce the sustainability challenges related to the fur product supply chain, such as the zero-waste principle, environmental impact, and ethical concerns. We then discuss sustainability and present the industry's political and economic impact. Next, we present the design, execution, and findings of our interviews with fur industry representatives. The interviews served as a means of evaluating the consistency of our findings from the secondary data. They also resulted in new findings that were beyond the scope of the secondary data, such as industry outlook. We then introduce and discuss a holistic model of fur industry sustainability. Lastly, we present this study's conclusions.

Brief Introduction to the Fur Industry

The fur industry is ancient. Its original purpose was to protect wearers from extremely cold weather. However, fur coats have also been worn in warmer regions as a symbol of wealth and power. Historically, because fur came solely from trapped wild animals, it was difficult to make a perfect fur coat without any of the natural blemishes incurred during the animal's lifetime. Thus, a perfect fur coat was a precious item, and because fur coats are durable, they were typically passed on to subsequent generations. Today's well-organized fur industry primarily uses the fur farming method, which accounts for approximately 85% of fur production, whereas the secondary trapping method accounts for approximately 15% of production (TRUTH, 2017).

In fur farming, animals such as mink and fox are farmed in a similar manner as cows, chickens, and pigs because of soon-to-be-implemented regulations on animal welfare standards with which fur farms must comply in order to qualify for certification (Fur Europe, 2017a). Additionally, farmed furbearing animals are well-treated to ensure highly uniform quality of growth, which consequently ensures the quality of the fur's color (i.e., minimal variation in quality). Perhaps an appropriate comparison would be to how Japan's Kobe cows are raised.

Trapping of wild furbearing animals is practiced in certain countries located near the Arctic Circle, such as Canada and the USA. Unlike traditional hunting, trapping devices are specially designed to eliminate any possible damage to the animal's skin and bone. As stated, high-priced fur products require perfect fur pelts, and any natural damage to the fur is undesirable. For the sake of balancing the ecosystem, local governments regulate the species and the number of animals that they allow to be trapped. These wild animals cannot be farmed.

The fur industry is controversial and subject to criticism and attacks by antifur cruelty pressure groups, such as the People for the Ethical Treatment of Animals (PETA, 2017a), the Association for the Protection of Furbearing Animals (FurBearer, 2017a), the International Anti-Fur Coalition (IAFC, 2017), the Anti-Fur Society (AFS, 2017a), and the Fur Free Alliance (FFA, 2017). These pressure groups demonstrate against the industry and advocate the prohibition of fur products. Some of these groups use news and advertising media to promote fashion design brands (FFA, 2017; FurBearer, 2017b) that have banned the use of natural fur materials in their products. They also promote synthetic alternatives to natural fur—namely faux fur—as a means of combating the fur industry (FurBearer, 2017c). Many of these groups request donations from their supporters (PETA, 2017a, AFS, 2017a). The Anti-Fur Society, a US-based group that does not restrict its political involvement, focuses on attacking the fur industry in China (AFS, 2017b). Hence, there is clearly much intense debate (TRUTH, 2017) and political lobbying targeted at the fur industry.

Despite the antifur movement, the fur industry grew rapidly from US$15 billion in 2011 to US$40 billion in 2015 (Kor, 2017). The supply of fur comes primarily from countries in Northern Europe (e.g., Denmark, Hungary, Romania, Finland, Sweden, Norway, and Russia), Western Europe (e.g., France, Spain, Italy, and Greece) (Fur Europe, 2017a), and North America (e.g., Canada and the USA). Some South American countries, such as Brazil, produce fur. China is also a major fur-producing country. Mr. Mark Oaten, CEO of the IFF, reported that Finland and China together accounted for close to 95% of the nearly nine million fox furs produced globally in 2015 (WearFur, 2017).

Fur pelts are largely traded in auction houses worldwide. European auction houses include Kopenhagen Fur (2017), Saga Fur (2017), and Sojuzpushnina (2017). North American auction houses include North American Fur Auction (NAFA, 2017), Fur Harvesters Auction Inc. (2017), and American Legend (2017). After the pelts are sold to manufacturers, the apparel makers undertake the production of fur coats and other products. The fur pelts are first put through processing, including dressing (tannery for fur skins), dyeing, and are then cut into patterns of specific shapes and sizes according to the product design. Unlike mass-produced apparel, fur pelts must be cut layer by layer individually, because the material is natural and thus its shape is irregular. The cut pattern pieces are then sewn into a fur coat or other product. Finally, after passing quality inspection, the products are sold to consumers. In addition to apparel manufacturers, wild fur might be distributed to fur dressers or private users. However, the quantities involved in these distribution channels are comparatively small.

The IFF estimates that the fur industry employs approximately one million people worldwide (IFF, 2017). Fur manufacturing typically occurs in communities where relevant skills are passed from generation to generation. The number of formal fur fashion design programs worldwide is very limited. For example, the Centria University of Applied Sciences of Finland offers a BA program in International Business, Fur Design and Marketing (Centria, 2017). However, in most other geographic areas, including China and Hong Kong, fur designers are trained as regular fashion designers who happen to use fur pelts as their medium. The IFF hosts its annual Remix International Student Fur Design Competition in Milan (FICA, 2017). Qualifying competitions are held elsewhere, such as the 2017 Fur Design Competition in Hong Kong (HKFF, 2017). Winners of the regional competitions can then participate in the international competition in Milan.

In summary, the global fur industry is both growing and highly organized. The industry provides employment and other economic contributions wherever it operates. The industry's most crucial sustainability challenges are environmental and ethical concerns, which constantly appear on the agendas of antifur pressure groups. However, the political and economic impact of antifur activism on the fur industry has been limited because many of the fashion brands that ban the use of natural fur do not offer any regular fur product lines, and some lack market share in fur-selling countries. Hence, the key players in determining fur industry sustainability are the people who make their living in the fur product . It is therefore logical to interview representatives of fur industry associations, leading fur enterprises, and fur product designers. Fur industry associations play a role in negotiating with governments on various political, economic, and environmental problems related to the fur industry. Fur-producing enterprises and fur designers support and execute these agreements. They also play major roles in product innovation, supply chain improvement, and new market exploration.

Fur as a Sustainable Luxury Product

In this section, we present arguments in favor of two concepts: (1) fur products are luxury items, and (2) the demand for sustainable fur products is increasing. In the subsequent section, we continue with a discussion of fur products' sustainability.

Fur Products as Luxury Items

Fur products are expensive in terms of retail prices. However, high-priced products are not considered to be luxuries by default. As Magee (2016) described in her study of the meaning of luxury for fur products in Poland, she proposed the definition of luxury as "an excess of normal". Clearly, with today's technology, there are alternatives to fur for maintaining one's body warmth. Hence, it is logical to discuss fur products as both excessive and expensive—and thus as luxury products.

The demand for fur products has gradually increased along with the number of affluent consumers. Their desire for fur products can be explained by fashion and cultural theories. According to fashion theory, the high price of fur products has caused them to become a symbol of both affluence and the fulfillment of consumers' desires (Barnard, 2007). The quest for individual identity is an inherent human desire, and by wearing fur products amid the context of "excess," wealthy consumers can differentiate themselves from those who are less privileged and can only afford industrialized, machine-made products such as faux fur. This attitude was also observed in the Victorian era (Bellatory, 2017), during which handcrafted products were a symbol of economic status because machine-made products had become affordable due to industrialization. Crucially, consumer demand for fur is both undeniable and sustainable, as strongly evidenced by PETA's Web page (PETA, 2017b), which introduces fur substitutes to fulfill consumers' desire for fur. If this desire could be eliminated, there would be no need to introduce faux fur as a replacement for natural fur.

Cultural theory suggests that luxury items are commonly passed down from generation to generation, both as a means of expressing love and because of the items' resale value (Niinimäki, 2015). Jewelry, fur coats, and kimonos are examples of such items. The life span of fur products is very long; a properly maintained fur coat can easily last for 50 years or more. In many cultures, such as those of northern China, Finland, and the USA, it is common for mothers to pass their most beloved garments on to their daughters (Niinimäki, 2015).

Furthermore, manufacturing fur products is very complicated and requires skilled labor. In many countries, such as Canada, Finland, and Denmark, fur production is a part of their national heritage. Citizens of such nations take pride in preserving these skills, not only because they use them to make a living, but also because they are part of their heritage (Dragon, 2017). For many people, the desire for fur products is inherent in their culture.

Fur Products as Sustainable Products

Different people may have divergent understandings of sustainable products because the concept of sustainability varies from culture to culture and from product to product. This section discusses sustainability in terms of the supply chain, including (1) sources of raw materials, (2) processing of raw materials and ethical considerations, (3) controlling the chemicals used in processing, (4) handling of skins and by-products, (5) consumer protection, (6) product life span, and (7) environmental considerations. We further discuss the global political and economic impact of the fur industry to support our argument that the fur industry is sustainable.

The two major methods of collecting raw materials in the industry are furbearing animal farming and trapping. Historically, all fur products were produced by trapping and hunting wild animals. Today, luxury fur products are so expensive that it makes sense to control the supply of raw materials. One method of doing this is to rely on fur farming, which allows for the flexibility to increase or decrease the fur supply according to predicted demand. Another method is to rely on wild animal hunting and trapping, which is less flexible in terms of supply because of legal restrictions on the number of wild furbearing animals that can be hunted. However, scarcity has become a justification for high prices. As a result, more than 85% of fur industry production is currently based on fur farming. The remaining 15% of fur industry production depends on which species require culling in order to maintain an ecological balance (Skov, 2005).

The most effective method of assuring fur quality is to raise an entire generation of animals in the same environment using the same quality of food, so as to minimize differences in skin quality among individual animals. This is critical because many fur products are produced using skins from two or more animals. High food quality and appropriate daily exercise are important to produce quality skin and fur. Furthermore, the furbearing animal farms help to reduce overall food wastage by feeding furbearing animals with unused parts of other non-furbearing animals such as chicken bones that are crashed and mixed into the feeds.

When the animals grow to a predefined age or weight, they are terminated with minimal pain. The Trade Association Dansk Fashion and Textile's code of conduct states the following: "in businesses where animals are used in labor and/or in the production (fur, wool, etc.), such animals must be fed; treated with dignity; respected; and no animal must deliberately be harmed or exposed to pain in their life span" [Kruger et al., 2012, 135 (cross-reference Niinimäki, 2015)]. The remains of the furbearing animals are converted into fertilizer or food for other animals.

Among the many animal welfare awareness programs, WelFur is a farm-level animal welfare certification program that focuses on reliably assessing and improving animal welfare (WelFur, 2017). The program serves as a scientific reference for the regulation and control of European fur farms. The European Commission considers WelFur as a supporting reference for the development of a pan-European animal welfare framework law. The European Commission's Welfare Quality project defined 12 criteria and included many relevant measures up to the species-specific level (Welfare, 2017). Furthermore, the principles of the Welfare Quality project are comprehensive and include proper housing, adequate feeding, robust health, and appropriate farm worker behavior. North America also has many regional certification programs, the details of which vary by the target wild animals and by the state or province. However, these programs all share a common goal, which is to inform consumers that furbearing animals are treated in accordance with industry standards. All these programs fall collectively under the FurMark program (FurMark, 2017).

The skins are shipped to fur dressers for further processing. This is the chemical processing stage, which involves dyeing the skins and other finishing processes. Similar to the chemical processes of other raw materials for the apparel industry, dressers must implement environmental controls on the chemicals used and engage in research and development on environmentally friendly chemicals and processes (IFDD, 2017).

The majority of processed skins are shipped to factories to manufacture fur apparel. Very few skins are shipped to fur dressers or private customers. The fur apparel production process is very different from that of other types of apparel. For example, the cutting pattern of a fur garment panel is based on an individual animal and therefore cannot be handled by computer-aided design systems. However, cutting wastage is practically zero because the smaller pieces are used to manufacture fur accessories. The meat is sold to farms as food for other animals, and the bones are converted to fertilizers (DSS Management Consultants Ltd., 2011). Thus, this manufacturing process is practically zero-waste.

Because all retailers must attract consumers to return and purchase again, a traceability program, Origin Assured (OA), has been instituted by the IFF (FCC, 2017; Pavuna, 2017) to assure consumers of a product's quality and origin. Every OA-certified product can be traced back both to the garment manufacturer and the relevant furbearing animal farm or farms. This program ensures product quality and the ethical standards of furbearing animal farms.

The life span of fur products is very long. As stated earlier, based on the cultural theory argument, the life span of fur products can be extended to multiple generations. Moreover, the life cycle analysis undertaken by DSS revealed that based on the experience of furriers, an estimated 10% of the fur used in natural fur garments can be reused even after the useful life of the original fur garment has ended (DSS, 2011). From this perspective, fur products are clearly more environmentally friendly than fast fashion products that are manufactured using non-biodegradable man-made fibers.

Fur products are environmentally friendly in the sense that real fur is both organic and biodegradable. As the environmental friendliness of the dyestuff and other chemicals that are used in the processing of fur continues to improve as technology advances, fur products will benefit from higher demand. Because antifur pressure groups frequently claim that fur products are not environmentally friendly and promote the use of faux fur, it is appropriate to investigate these claims. Based on DSS's comparative life cycle analysis of natural and faux fur, the former is more environmentally friendly in the following ways: (1) Natural fur is both organic and biodegradable, whereas faux fur is non-biodegradable because it is a by-product of petroleum; (2) mink farming can produce by-products such as nitrogen and phosphorus that can help the environment, whereas faux fur cannot; (3) the manufacturing of a faux fur coat poses a much greater risk of potential climatic impact; (4) the manufacturing of a faux fur coat can potentially require more than three times the resource consumption of a real fur coat; (5) the lower cost of faux fur might induce higher consumption, implying a heavier load on the supply chain, heavier carbon dioxide emissions, and more wastage of faux fur products (DSS, 2011; Olson & Goodnight, 1994; Skov, 2005) (cross-reference Reed, 1989). Table 8.1 summarizes the raw scores that DSS used to compare the performance of natural fur and faux fur. The raw score reflects the negative impact of natural and faux fur on various environmental concerns. A higher value reflects a greater negative impact. These figures could be informative for consumers (DSS, 2011).

The Danish Agriculture and Food Council (DAFC, 2017a) explained the environmental impacts of fur farming, including the reduction of CO2-equivalent greenhouse gasses from 18.7 million tons in 1990 to 13.4 million tons in 2008 in

Table 8.1 Life cycle scores and percentage differences for individual endpoint indicators (DSS, 2011)

Endpoint impact category	Raw score		Difference (%)
	Natural fur	Faux fur	
Human health	95.72	98.80	3
Ecosystem quality	−11.87	23.83	−301
Climate change	49.58	113.45	129
Resources	58.43	157.02	169

Denmark. Given that the mink population is approximately three million and the total population of livestock in Denmark is approximately 45 million, the number of fur farms is approximately 6.7% of the total number of Danish farms, meaning that the contribution is quite substantial (DAFC, 2017b).

Every year in Finland, thousands of animals must be hunted to balance the ecosystem. The fur from these animals is sold to designers such as Huurinainen (2017), who has created a collection known as WILD Concept, the name of which signifies that the fur used came from animals that spent their lives in the wild. In this sense, "fur consumption is beneficial to animal populations and the environment" (Olson & Goodnight, 1994). Similar practices can be found in Canada, the USA, Russia, and elsewhere.

Thus, fur products are clearly sustainable products because they are more environmentally friendly than faux fur along the full supply chain. Next, we discuss sustainable aspects of the fur industry and its political and economic impact on the world.

Fur Industry as a Sustainable Industry

Having undertaken a micro-analysis of fur products in the preceding sections, we now present a macro-analysis of the fur industry. This industry impacts several aspects of the global political and economic environment, including through (1) its political advocacy and (2) its contributions to trade and employment.

In its political activities, the fur industry advocates on policies related to animal welfare and consumer protection. The FurMark program, which includes WelFur, is an example of government policy to promote the ethical treatment of animals. OA is an example of an initiative promoting consumers' "right to know." The fur industry also strongly supports the protection of endangered species.

The fur industry plays a key economic role in creating jobs and contributing to gross national product in the countries in which it operates. The IFF estimated full-time employees in the industry at more than one million, with many more employed on a part-time basis (IFF, 2017). Fur Europe estimated that more than 5000 fur farms operate in 22 European countries (Fur Europe, 2017a). Furthermore, the Fur Information Council of America estimated that there are 1100 retailers and 100 manufacturers (FUR, 2017a). Turkistieto (2017) stated that there were 973 fur farms in Finland in 2017, and industry employment in Finland was approximately 4300 person-years in 2012, according to Pellervo Economic Research. The Fur Institute of Canada (FIC, 2017) estimated that there are 50,000 active trappers and 289 operating farms in Canada. In addition, the DAFC (2017c) reported 1500 active industry farmers in Denmark in 2017. These figures refer to direct employment, including both full-time and part-time employees, but they do not reflect indirect employment in the industry's supporting infrastructures, such as logistics, banking, retailing, and agriculture. Therefore, assuming demand for fur products remains sustainable, the fur industry should also be sustainable.

Global production of unprocessed fur pelts amounted to approximately US$7.8 billion in 2012–2013 (Hansen, 2014). Moreover, PwC Italy has estimated retail fur sales to be US$40 billion (IFF, 2017). The Fur Information Council of America (FUR, 2017) reported US$1.5 billion in US sales in 2014 following a 7.3% annual increase, and the FIC (2017) reported sales of CAD$1.0 billion in Canada in 2013. Furthermore, the DAFC (2017d) reported that according to the Kopenhagen Fur Center, sales reached EUR1.1 billion in 2015, accounting for 7% of Denmark's total exports.

The preceding estimates indicate the fur industry's direct economic contribution, both globally and in the specific nations discussed. Indirect employment and economic activity are also induced in support of this trade. This further underscores the argument that the fur industry is both powerful and sustainable. The industry has been experiencing strong growth in worldwide aggregate sales, from US$15 billion in 2011 to US$40 billion in 2015. We expect fur product sales to continue increasing as the world economy becomes further polarized between the rich and poor [as measured by the Gini coefficient, which has increased from 0.317 in 2007 to 0.318 in 2014 or the latest (OECD, 2017)] and as China actively promotes its "Belt and Road Initiative." China is the world's largest importer of fur, accounting for approximately 50% of global fur imports (Hansen, 2014). Moreover, the initiative will also boost economic growth in all countries along the Silk Road, including Mongolia, Kazakhstan, Ukraine, and Russia (HKTDC, 2017). This should result in increased demand for fur products because these countries are located in the polar and temperate zones.

We have presented an objective picture of the fur industry and have used figures from various reliable sources to argue that the industry is sustainable in terms of its demand, product, and industry outlook. However, do people who are working the fur industry have a similar view on the industry? Are there different perspectives that lie beyond these objective figures? In the next section, we will present the results of a series of interviews with employees in various sectors of the fur industry.

Interviews

After compiling the objective data and arguing in favor of the sustainability of fur demand, product, and industry outlook, we conducted interviews with industry employees to identify their subjective evaluation of fur industry sustainability. We consider interviewing to be a more appropriate method than issuing broad surveys because experts and industry leaders have access to undisclosed information and will determine the future of the industry. It was our honor and pleasure to interview Mr. O, CEO of a global fur-related association; Ms. X, CEO of a regional fur-related association; Mr. F, CEO of a global fur brand; and Ms. I, senior manager of a fur auction house. Because the fur auction house assists designers and manufacturers to build business relationships, Ms. I is highly familiar with the latest developments in design trends and innovation (Nexus, 2017). The interview was conducted on February 7, 2017, in the Hong Kong Convention Centre, during the 2017 Hong Kong International Fur and Fashion Fair. Table 8.2 lists the interview questions.

Table 8.3 illustrates the encoded identities of the interviewees, and Table 8.4 summarizes and paraphrases the outcome of the interviews. When more than one interviewee shared similar ideas, their opinions are summarized and credited together. The order of presentation follows the same order of different levels of sustainability, namely sustainability of demand of fur product, sustainability of the fur product, sustainability of the fur industry.

The interview results exhibit a high level of coherence with our observations from the secondary data, but they also offer new perspectives. In particular, the fur industry's strategy and outlook cannot be derived from the secondary data, whereas the interview results reflect the subjective evaluation of both current trends and predictions for the near future. They also illustrate the strategies that are the foundations of global fur industry sustainability.

Holistic Model of Fur Industry Sustainability

Analyzing the sustainability of the fur industry is more complicated than assessing the sustainability of other apparel manufacturing industries. Studies of the sustainability of the overall apparel industry have not included fur as a raw material. There is no realistic concern about shortages of raw materials, but global antifur pressure groups could block access to the raw materials. Therefore, an assessment of the supply chain is insufficient to analyze the sustainability of the fur industry. We propose a holistic model of the fur industry's sustainability issues focusing on two major areas: economics and politics. These areas involve the supply chain for luxury fur products. The supply chain is divided into three parts: supply, process, and demand. The sustainability of supply depends on both cultural and economic factors. The sustainability of the process depends on technological advancements. The sustainability of demand depends on cultural and psychological aspects where

Table 8.2 List of interview questions

	Interview questions	Related to sustainability of
1.	People today are talking about green fashion, organic fashion, and sustainable fashion. In the future, what role will the fur industry play in developing green, organic, and sustainable fashion?	Product
2.	Why is fur often considered to be a green product in the industry? What is the fur industry doing to raise its standards?	Product
3.	What are the waste products of the fur apparel manufacturing process? How do they impact the environment?	Product
4.	Other than environmental aspects, what economic and social benefits does the fur industry provide?	Industry
5.	Fur products have long been considered luxuries. Assuming even greater future improvements in farmed animals' treatment and environment, higher-quality materials, more innovative designs, and more scarce supply, fur products should become even more luxurious, and profit margins should therefore dramatically increase. Increased profit margins imply more concentrated supply and production, in which case the killing rate might decrease. Do you agree? Do you think it is possible that fur products might become "ultra-luxurious" to increase profit margins and maintain the sustainability of the industry?	Demand
6.	Your firm has an in-house research team. How creative do you see fur product design becoming in the future?	Product
7.	Do you think fur is more of a functional product than a fashion product? Or do you think it is the other way around?	Demand, product
8.	How does your auction house facilitate the activities of other fashion houses and designers?	Demand
9.	How do you ensure buyers' and users' confidence and comfort in the fur products they purchase and wear?	Product
10.	Do you consider fur to be a green product? Why? What role does fur play in sustainable fashion? What is the fur industry doing to support environmental conservation?	Product
11.	Have you seen any changes in fur consumption trends in recent years?	Demand
12.	Is a mutual global standard for animal welfare protection important?	Product, industry
13.	How do faux and natural fur differ when you use them to tell your design stories?	Demand
14.	Where do you see the fur industry in 10 years?	Product, industry

Table 8.3 Coding of interviewees

Code of interviewee	Interviewee
O	Mr. O, CEO of global fur-related association
X	Ms. X, CEO of regional fur-related association
F	Mr. F, CEO of global fur brand
I	Ms. I, senior manager of fur auction house

Table 8.4 Themes emerged from interviews

Topic	Themes emerged from interviews	Group of interviewee(s) making similar comments
Sustainable demand for fur products as luxury products	1. Need for new product development: New products and markets can objectively ensure the sustainability of demand for fur products. Fur product consumption is growing in the youth segment. Fashion brands can imbue their product lines with a more prestigious and luxurious image by adopting fur as a design element. Furthermore, new low-density fur products have been produced for consumption in regions where it is too hot to wear traditional fur products. These products use grid- and net-based fabric to increase air ventilation and heat flow. Fur products that are manufactured using this type of fabric can be worn in air-conditioned environments such as hotel ballrooms anywhere in the world	I
	2. Coexisting with faux fur: Designers use faux fur to imitate real fur in an attempt to tell the same story of social status. This only confirms that consumers desire to be perceived as affluent. Real fur is the status quo for the more prestigious social classes	I
	3. Fur as heritage products: With more than two thousand manufacturing techniques, the fur industry is an important heritage of Europe. Many European workers are proud to continue to develop this highly skilled art form	I
	4. Fur as functional products: Many places have special fur societies and communities. This is especially true of countries where fur materials come from isolated, extreme weather environments, whose inhabitants rely on fur to sustain their everyday lives. Demand for fur in these communities is clearly sustainable	O, X, F
Fur products as sustainable products	1. Sustainable manufacturing processes: The IFF is very conscious of the sustainability of the entire manufacturing process from start to end (e.g., farming, dressing, and dyeing). The fur industry already has a lead in this area, resulting in a strong argument for its overall sustainability. For example, conservation is an important part of environmental protection during the manufacturing process. The industry has been working actively in conservation (e.g., preserving marine and forest life) to maintain the supply of fur and support the welfare of certain furbearing animal species. The IFF supports the International Union for Conservation of Nature and employs experts on its conservation team. Both the natural fur product and the elements surrounding fur should be sustainable. Real fur is both biodegradable and organic, whereas faux fur is neither.	X, O, F

(continued)

Table 8.4 (continued)

Topic	Themes emerged from interviews	Group of interviewee(s) making similar comments
	Chemicals used in the production of both real fur and faux fur require more innovation to be more environmentally friendly. However, the fur industry is working on ensuring that all materials used in the value chain are natural and that the value chain is organized effectively. As the technology improves and certification of animal welfare and environmental protection standards becomes more widespread, both demand for and supply of fur products should be further strengthened	
	2. Increasing transparency of the supply chain: Certification programs for consumer protection and the control of unethical furbearing animal farming and trapping are crucial. IFF continues to work with its partners to create global standards, with clearer labeling and messaging to brands and consumers. Establishing a worldwide certification and traceability program is complicated, but IFF is committed to bringing this together as its top priority. Saga Furs has its own supplier certification and auditing programs that monitor suppliers' animal welfare and environmental practices. Saga Furs' mink fur is fully certified. The entire industry is taking initiatives to move forward in Europe with WelFur, because it offers a scientific way to measure and control these practices. WelFur will also be a part of the Saga certification system to measure animal welfare. To ensure the credibility of the certification, independent auditors monitor the entire program	O, F
	3. Creating and preserving local communities: In countries such as Denmark and Finland, the fur industry is one of the main creators of direct industry and indirect supply chain jobs	O, F, I
	The fur industry will continue to prosper, grow, and innovate. Growth trends are strong in product variety and market segment expansion. The industry will focus on the youth market by collaborating with designers to innovate more in order to increase young consumer demand. Mixing and matching of design styles and integrating fur elements with other materials should further extend fur products' potential	O, F, I
Outlook of the fur industry in 10 years	New technological innovations must be developed. The greatest challenge ahead is to establish a clear traceability and certification program	O
	The IFF will continue to support efforts to protect animal welfare and the global environment	O, F, I

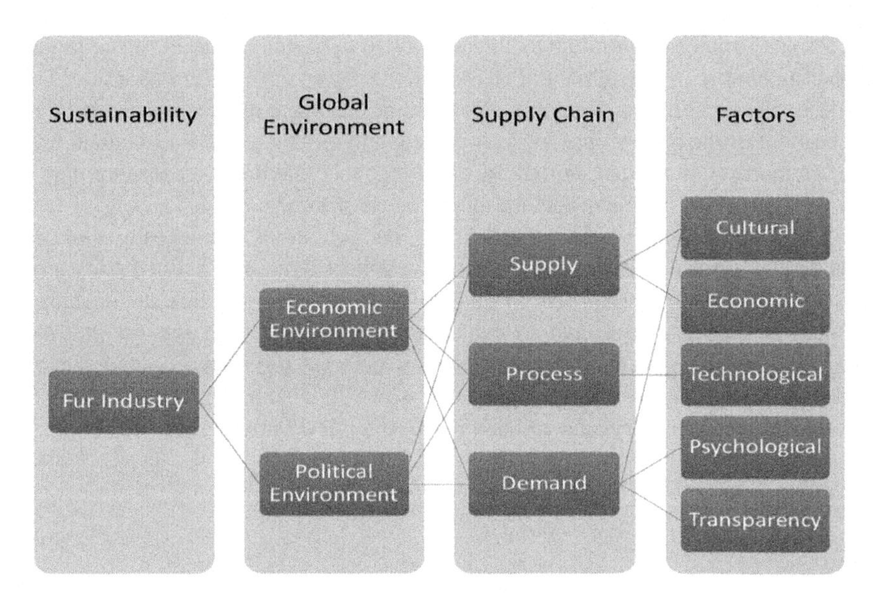

Fig. 8.1 Holistic model of the fur industry's sustainability issues emerged from the study data

transparency is playing an important role of ensuring the consumers' confidence in the well-being of the furbearing animals. Figure 8.1 outlines the model.

Based on objective and subjective evaluations, the sustainability of the fur industry is certain despite continued attacks from antifur pressure groups. The environmental arguments put forth by antifur activists are often equally applicable to faux fur production. Given that the world is searching for greener processes, the fur industry should be able to increase its scale.

Although antifur pressure groups' ethical arguments tend to be equally applicable to other animal farming industries, fur has a distinctive feature: It symbolizes affluence and social status. Could the psychological response of hating the affluent be the fundamental force behind the antifur movement? Because psychological analysis is beyond our scope, we propose that the fur industry launches a separate study of this question.

The ethical concerns raised by killing furbearing animals could be addressed through the power of religion. For example, kosher slaughter (shechitah) is a Jewish ritual involving the slaughter of kosher species without causing any pain (KIR, 2017). Halal slaughter is an Islamic ritual that requires the slaughterer to pray beforehand and is also carried out in a painless manner (IFANCA, 2017). Buddhists pray to help the spirits of slaughtered animals to be reborn in the next life (Buddhist, 2017). Because it remains unclear whether these additional religious elements might result in the approval of fur products among antifur consumers, we propose that the fur industry conducts marketing research on this topic.

Lastly, a simple solution to the ethical problem of farming is to allow the farmed furbearing animals to die of natural causes such as aging prior to fur extraction. This would be the most humane manner of obtaining fur, but the economic costs might rise because of the longer time frame involved, and the quality of fur obtained from older animals might not be optimal. It is also arguable that the extra feeding may or may not contribute to the sustainability of the environment.

Finally, we must acknowledge and discuss the limitations of this study. Our proposed model is based on the information that we collected in carrying out our research. The objective of this study was to undertake preliminary research, to develop a vigorous academic model, and to explain the sustainability of the fur industry. Therefore, we attempted to identify the relevant factors that should be included in our model. However, it is possible to discuss the industry's sustainability based solely on the quantitative secondary data and qualitative interview data.

Conclusion

Our research consisted of an investigation of the sustainability of the fur industry. The first level of sustainability is demand for fur products, the next level is the environmental impact of fur production, and the final level is the fur industry's global economic impact. We conducted the study in two phases: an objective review of the secondary data and a subjective evaluation of industry sustainability obtained by interviewing representative industry leaders. The objective data serves to support the findings of our subjective evaluation. At the same time, the interview results offer new perspectives that go beyond the objective data, particularly when assessing the industry's strategy and short-term outlook.

The sustainability of fur product demand is strong largely because fur is a symbol of luxury. Because it is no longer necessary for protection from the cold and commands a high price, fur signals the affluence of its wearers. The success of faux fur provides solid support for the argument that desire for fur is strong. Furthermore, the fur industry is expanding product lines to target youth markets and creating new designs so that fur products can be mixed and matched more effectively with non-fur products. Innovative low-density fur designs have resulted in fur products that can be worn comfortably in air-conditioned environments anywhere, thus extending potential markets for fur products to warmer parts of the world. The desire for fur products comes not only from consumers outside the industry, but also from the suppliers themselves. Many European fur industry workers take pride in maintaining this aspect of their cultural heritage, because fur production is a highly skilled and complex art form. Specialist fur societies and fur communities depend on the fur industry for their survival. Consequently, the demand for fur is sustainable.

Fur products are also environmentally sustainable. Clearly and objectively, fur comes from animals, so natural fur is both organic and biodegradable, unlike faux fur. Although some chemicals, such as dyestuff, that are used in the processing of fur are not entirely environmentally friendly, they are registered in the REACH system (EU, 2007). As technology advances, more environmentally friendly manufacturing methods will replace previous ones. Moreover, similar problems exist in the production of all industrial consumer products, including faux fur. Moreover, fur farming utilizes a zero-waste approach within the entire supply chain, in the sense that all parts of the farmed animals are utilized in sectors such as apparel, food, and agriculture. Furbearing animal trapping methods are a means of annually culling excess populations of animals to maintain ecological balance. Hence, fur products play an important role in environmental conservation. Furthermore, fur industry leaders are fully aware of their responsibilities in environmental protection and the ethical treatment of animals. For example, the IFF strongly supports the International Union for Conservation of Nature. The fur industry relies on the WelFur program for certification of fur farms in order to meet the required animal welfare standards of Europe, and it relies on the FurMark program for similar purposes in the USA and Canada. Together with the OA traceability program, these programs can assure consumers that the fur products they purchase come from well-regulated sources and that their manufacturers have fulfilled the industry's ethical standards.

The fur industry is sustainable because of its political and economic impacts. Despite the activities of antifur pressure groups in some regions, global sales of fur products increased from US$15 billion in 2011 to US$40 billion in 2015. This implies that the fur industry contributes substantially to employment, with the latest estimate of one million people currently working full-time or part-time in the global fur industry. Moreover, the fur industry contributes to tax revenues and creates jobs and profits in related and supporting industries, such as banking, logistics, food, and agriculture. A major reason for this impressive performance is increased demand as more people become affluent enough to afford fur products. We expect the Belt and Road Initiative to increase sales of fur in countries along the Road as their citizens receive its benefits. For these reasons, fur industry leaders are confident about their industry's future and expect prosperous years ahead.

Acknowledgements We would like to express our gratitude and appreciation to the interviewees for their assistance and sharing of opinions: Mr. Mark Oaten, CEO of the IFF; Ms. Kelly Xu, CEO of International Fur Federation Asia Region; Mr. Pertti Fallenius, CEO of Saga Furs; and Ms. Julie Iversen, Head of Kopenhagen Nexus of Copenhagen Fur. We would also like to express our gratitude and appreciation to Mr. Robert Cahill, Senior Vice President of North American Fur Auctions, for his comments and suggestions. Last but not least, we would like to thank Ms. Abe Jong of The Hong Kong Polytechnic University who conducted and compiled the interview.

References

AFS. (2017a). http://www.antifursociety.org, visited June 1, 2017.
AFS. (2017b). http://www.antifursociety.org/Who_we_are.html, visited June 1, 2017.
AL. (2017). http://www.americanlegend.com, visited June 1 2017.
Barnard, M. (2007). Fashion theory: A Reader. In A. Barnard (Ed.), Routledge.

Bellatory (2017). https://bellatory.com/fashion-industry/Fashion-History-Victorian-Costume-and-Design-Trends-1837-1900-With-Pictures, visited June 1, 2017.
Buddhist. (2017). https://buddhists.org/buddhist-symbols/animal-death-beliefs-and-rituals-in-buddhism/, visited June 1, 2017.
Centria. (2017). https://web.centria.fi/applicants/degree-seeking-students/international-business-fur-design-marketing, visited June 1, 2017.
DAFC. (2017a). http://www.agricultureandfood.dk/prices-statistics/environment, visited June 1, 2017.
DAFC. (2017b). Facts & figures, denmark—A food and farming country 2016, p. 49, (http://www.agricultureandfood.dk/prices-statistics/annual-statistics, visited June 1, 2017).
DAFC. (2017c). http://www.agricultureandfood.dk/danish-agriculture-and-food/mink-and-fur, visited June 1, 2017.
DAFC. (2017d). Facts & Figures,Denmark—A food and farming country 2016, p. 44, (http://www.agricultureandfood.dk/prices-statistics/annual-statistics, visited June 1, 2017).
Dragon, B. (2017). Heritage fur products to warm the body and soul, http://www.truthaboutfur.com/blog/heritage-fur-products-warm-body-soul/, visited June 1, 2017.
DSS. (2011). A comparative life cycle analysis: Natural fur and faux fur, DSS Management Consultants Inc.
EU. (2007). https://ec.europa.eu/growth/sectors/chemicals/reach_en, visited 1st June 2017.
FCC. (2017). http://www.furcouncil.com/originassuredfur.aspx (mentioned OA™ via Originassured, 2017), visited June 1, 2017.
FFA. (2017). http://www.furfreealliance.com, visited June 1, 2017.
FHA. (2017). https://www.furharvesters.com, visited June 1, 2017.
FIC. (2017). http://fur.ca/fur-trade/canadas-fur-trade-fact-figures/, visited June 1, 2017.
FICA. (2017). Remix international fur design competition, https://www.fur.org/20172018-iff-remix-international-student-fur-design-competition/, visited June 1, 2017.
Fur Europe. (2017a). http://www.fureurope.eu/fur-information-center/fur-industry-by-country/, visited June 1, 2017.
Fur Europe. (2017b). http://iffti2018.csp.escience.cn/dct/page/1, visited June 1, 2017.
FUR. (2017). http://www.fur.org/fica-facts/, visited June 1, 2017.
FurBearer. (2017a). http://thefurbearers.com, visited June 1, 2017.
FurBearer. (2017b). http://thefurbearers.com/the-issues/fashion/fur-free-designers, visited June 1, 2017.
FurBearer. (2017c). http://thefurbearers.com/the-issues/fashion/alternatives-to-using-fur, https://www.fjallraven.com/responsibility/nature-environment/animal-welfare, https://varuste.net/en/Fj%E4llr%E4ven+Arctic+Fur?_tu=43463, visited 1st June 2017.
FurMark. (2017). FurMark—The international mark of responsible fur, International Fur Federation.
Hansen, H. O. (2014). The global fur industry: Trends, globalization and specialization. *Journal of Agricultural Science and Technology*. A, 4(7A).
HKFF. (2017). Fur Gala 2017, http://www.hkff.org/furGala.php?lang=1, visited June 1, 2017.
HKTDC. (2017). The belt and road initiative: Country profiles, http://china-trade-research.hktdc.com/business-news/article/One-Belt-One-Road/The-Belt-and-Road-Initiative-Country-Profiles/obor/en/1/1X000000/1X0A36I0.htm, visited June 1, 2017.
Huurinainen, M. (2017). http://ma-ri-ta.com, visited June 1, 2017.
IAFC. (2017). https://www.antifurcoalition.org, visited June 1, 2017.
IFANCA. (2017). http://www.ifanca.org/Pages/index.aspx, visited June 1, 2017.
IFDD. (2017). http://www.internationalfur.ca/index.html, visited June 1, 2017.
IFF. (2017). http://www.businessoffur.com/fur-info/industry-information/, visited June 1, 2017.
KF. (2017). http://www.kopenhagenfur.com/auction, visited June 1, 2017.
KIR. (2017). http://www.koshercertification.org.uk/whatdoe.html, visited June 1, 2017.
Kor. (2017). The International Fur Trade: Data and Measurements, https://dnnsociety.org/2016/12/05/international-fur-tradedata-and-measures/, visited June 1, 2017.

Kozar, J. M., & Hiller Connell, K. Y. (2015). Measuring and communicating apparel sustainability, Woodhead Publishing Series in Textiles (pp. 219–231). Amsterdam: Elsevier

Kozlowski, A., Searcy, C., & Bardecki, M. (2015). Corporate sustainability reporting in the apparel industry: An analysis of indicators disclosed. *International Journal of Productivity and Performance Management, 64*(3), 377–397.

Kruger, H., Himmestrup Dahl, E., Hjort, T., & Planthinn, D. (2012). Guide lines II: a handbook on sustainability in fashion. Copenhagen: Sustainable Solution Design Association, SSDA.

Lakshmanan, A., Jose, S., & Chakraborty, S. (2016). Luxury hair fibers for fashion industry, Springer Science + Business Media Singapore 1. In S. S. Muthu & M. A. Gardetti (Eds.), *Sustainable fibres for fashion industry, environmental footprints and eco-design of products and processes* (p. 1–38).

Magee, S. (2016). An 'excess of the normal': Luxury and difference in Polish fur critique. *Journal of Material Culture, 21*(3), 277–295.

Islam S. U. & Mohammad, F. (2016). Sustainable natural fibres from animals, plants and agroindustrial wastes—An overview. Springer Science + Business Media Singapore 31. In S. S. Muthu & M. A. Gardetti (Eds.), *sustainable fibres for fashion industry, environmental footprints and eco-design of products and processes* (pp. 31–44).

Møller, S. H. (2008, August). Sustainability in mink production—A management perspective. In IX *international scientific congress in fur animal production—Halifax* (pp. 19–23), Canada: Nova Scotia.

NAFA. (2017). http://www.nafa.ca/auctions-2/, visited June 1, 2017.

Nexus. (2017). http://www.kopenhagenfur.com/news?tag=Kopenhagen%20Nexus, visited June 1, 2017.

Niinimäki, K (Ed.). (2013). Sustainable fashion: New approaches. Helsinki: Aalto ARTS Books. https://aaltodoc.aalto.fi/handle/123456789/13769.

Niinimäki, K. (2015). Ethical foundations in sustainable fashion. *Textiles and Clothing Sustainability, 1*(1), 3.

OECD. (2017). OECD inequality update 2016, Table 1, http://www.oecd.org/social/OECD2016-Income-Inequality-Update.pdf, visited June 1, 2017.

Olson, Kathryn M., & Thomas Goodnight, G. (1994). Entanglements of consumption, cruelty, privacy, and fashion: The social controversy over fur. *Quarterly Journal of Speech, 80*(3), 249–276.

Originassured. (2017). http://originassured.com (directed to International Fur Federation website), visited June 1, 2017.

Pavuna. (2017). https://pavunaparka.com (mentioned OA™ via Originassured, 2017), visited June 1, 2017.

PETA. (2017a). https://www.peta.org, visited June 1, 2017.

PETA. (2017b). https://www.peta.org/living/fashion/cruelty-free-clothing-guide/, visited June 1, 2017.

Reed, J. D. (1989). The furor over wearing furs. Time, December 18, 72.

Resta, B., Gaiardelli, P., Pinto, R., & Dotti, S. (2016). Enhancing environmental management in the textile sector: An organisational-life cycle assessment approach. *Journal of Cleaner Production, 135*(2016), 620–632.

Roos, S., Zamani, B., Sandin, G., Peters, G. M., & Svanstro, M. (2016). A life cycle assessment (LCA)-based approach to guiding an industry sector towards sustainability: the case of the Swedish apparel sector. *Journal of Cleaner Production, 133,* 691–700.

Routledge. (2014). Routledge handbook of sustainability and fashion. In K. Fletcher & M. Tham (Eds.,) *ProQuest ebook central*. Routledge: Taylor and Francis.

Saga. (2017). http://www.sagafurs.com/auction/, visited June 1, 2017.

Skov, L. (2005). The return of the fur coat: A commodity chain perspective. *Current Sociology, 53*(1), 9–32.

Sojuzpushnina. (2017). http://www.sojuzpushnina.ru/en/s/2/, visited June 1, 2017.

TRUTH. (2017). http://www.truthaboutfur.com, visited June 1, 2017.

Turkistieto. (2017). http://www.turkistieto.fi/Basic_Information, visited June 1, 2017.

WearFur. (2017). https://www.wearefur.com/new-production-figures-reveal-another-strong-year-global-fur-trade/, visited June 1, 2017.
Welfare. (2017). http://www.welfarequalitynetwork.net/network/44219/9/0/40, visited June 1, 2017.
WelFur. (2017). http://www.fureurope.eu/fur-information-center/facts-figures/, visited June 1, 2017.

Chapter 9
The Drivers and Barriers of Luxury Sector Retailers to Adopt Energy Efficiency Technologies in Hong Kong

Spencer S. C. Tao and Chris K. Y. Lo

Abstract In recent years, many fashion brands and retailers have been under pressure to achieve environment-friendly production and operations. To reduce carbon footprints, retailers are seeking ways to adopt energy efficiency technologies (EETs), to have better control of energy consumption through technology improvement or substitution. The application of EETs has been researched in developed countries, such as the European Union and the USA, in areas of public and mainly manufacturing sectors. Some large-scale general merchandise retailers, like Walmart, are also widely reported for their adoption of adopting ISO 14001 environment management system, and ISO 50001 energy management system, and many EETs. However, it is not clear how EETs are being implemented in luxury retailers, which tend to be smaller in size located in major metropolitan cities. The energy efficiency of retail shops in major fashion cities, such as New York, London, Paris, and Hong Kong (HK) is largely neglected in the literature. To explore how luxury sector retailers in major cities implement EET, we selected HK as our research context. HK, being one of the largest and competitive luxurious product markets (HK's all retail sales in 2016 was US$56 billion of which luxury products amounted about US$10 billion), has long been suffered from light pollution due to the spot lights on large billboards on buildings, and the window display of brick and mortar stores on the street. The lightings are always on that also brings serious concern on energy consumption. This chapter shall review the application of EETs, such as renewable energy technologies, sophisticated lighting devices, battery technologies, energy-saving devices, and smart energy management, in the luxury sector retail industry of HK. In this paper, we reviewed both drivers and barriers of the application of EETs in the HK's luxury retail sector. The discussions are based

S. S. C. Tao (✉)
Institution of Textile and Clothing, The Hong Kong Polytechnic University,
Room QT715, Kowloon, Hong Kong
e-mail: sctao@polyu.edu.hk

C. K. Y. Lo
Business Division, Institution of Textile and Clothing, The Hong Kong
Polytechnic University, Kowloon, Hong Kong
e-mail: tcclo@polyu.edu.hk

© Springer Nature Singapore Pte Ltd. 2018

C. K. Y. Lo and J. Ha-Brookshire (eds.), *Sustainability in Luxury Fashion Business*,
Springer Series in Fashion Business, https://doi.org/10.1007/978-981-10-8878-0_9

upon interviews of renowned luxury goods retail chain stores in HK. Six luxury and premium brands (i.e., jewelry, luxury watches, fashion) retail chain stores are chosen, and in-depth interviews were conducted with their senior managements, who are responsible for the decisions of EETs adoption (if any). The samples include both private and publicly listed companies, and their number of stores range from 5 to thousands retail outlets in HK and China. We believe the findings could be applied into other retailers in luxury sectors. A conceptual model, Energy Sustainability Strategy Model on EETs adoption, is also proposed.

Keywords Energy efficiency technologies · Institutional theory
Retailers

Introduction

Climate change has become a global challenge. As part of the international community, alongside 20 other Asia-Pacific Economic Cooperation (APEC) economies, HK targets to reduce energy intensity from the 2005 level by 40% by 2025. As a service-based economy, the major contributor of greenhouse gasses in HK is from electricity in commercial buildings. HK Government has launched various programs (e.g., Buildings Energy Efficiency Funding Schemes) to promote energy efficiency to building owners. However, current policies are constraints to building owners without direct implications or incentives to the business entities (often tenants) who are indeed responsible for the energy consumption decisions.

The retail industry in HK, its long operating hours, heavy use of lighting on the shop floor, windows display, outdoor advertising, and the intensive use of air-conditioning, consumes a large amount of energy in this sector. Therefore, energy efficiency is increasingly important for the retail industry to maintain its competitiveness. In HK, over 70% of the retailers are small and medium-sized enterprises (SMEs) who contribute a significant part of the gross domestic product. These SMEs, which comprise a significant proportion of service sector economic activity, have been overlooked by academics and policymaking as candidates for significant greenhouse gases' (GHG) emission reduction. Despite the economic importance and potential GHG emission reduction of SMEs in retail sectors, no prior study has investigated on how to improve the effectiveness of EETs adoption for retailing SMEs (particularly luxury products retailing SMEs) in HK.

Energy saving has long been one of the major issues on the topic of environmental sustainability. In the past, people were anxious about how the world resources could be better utilized as far as environmental sustainability is concerned (Abdelaziz, Saidur, & Mekhilef, 2011). With the advent of technology, we are now in another frontier to encourage using energy efficiency tools and renewable energy to prevent depriving of the world resources. Our objective is to study the drivers and barriers of how the luxury retail industry motivates to adopt the energy strategies in dealing with the environmental sustainability issues. We also try to understand the initiatives, challenges, and incentives in carrying out the energy sustainability strategies that lead to practical recommendations to retail industries in energy saving by and large.

In this research, we use the institutional theory as our theoretical lens to understand the energy-saving strategies of the HK's luxury retail stores. We also review the energy policies in HK and the energy aspect of sustainability in the luxury retail industry. We then propose a conceptual model of energy sustainability strategy followed by our identified drivers and barriers of implementing this model.

Background and Literature Review

A Review of Energy Policies in Hong Kong

HK renowned as a city of shopping paradise is one of the cities suffering from light pollution. The outdoor lighting is an inevitable by-product of a modern city, like HK, for attractive and decorative purposes. Pun, So, Leung, and Wong (2014) found that the overall night sky brightness of HK was about 100 times higher than the international standard of pristine night sky background.

The environmental policies in HK are governed by the Environmental Protection Bureau (EPB) of the HK Special Administrative Region (HKSAR) government. EPB issued an energy-saving policy (see Report "Energy Saving Plan for HK's Built Environment 2015 ~ 2025+") in 2015 with a key action to reduce energy intensity by 40% by 2025 from the base value set at 2005. This policy, with a slogan of being "Energy Aware" and "Energy Wise", laid down in detail the strategies and procedures of high energy saving pursued by the HKSAR government, and is driven by the combination of economics, regulatory, education, and social means to enhance HK's energy efficiency by 2025 through a strategy of public–private partnership (PPP). EPB will periodically review, expand, and/or tighten energy-related standards, namely Building Energy Efficiency Ordinance (BEEO), Building (Energy Efficiency) Regulation (BEER), and Energy Efficiency (Labeling of Products) Ordinance (EELPO). The EELPO comprises a Mandatory Energy Efficiency Labeling Scheme (MEELS) which sets out the practical guidance and technical requirements on energy efficiency labeling for compact fluorescent lamps, room air conditioners, refrigerating appliances, washing machines, and dehumidifiers.

This energy-saving policy is prioritized to focus on dealing with commercial and institutional (i.e., public sectors) buildings followed by residential buildings with an ultimate aim to achieve a locally defined building quality standard, called Building Environmental Assessment Method (BEAM). It is targeted to get building owners to adopt environmental design, energy-saving measures, and green building management. In addition, the HKSAR government also launched an Energy-Saving Charter on "No Incandescent Light Bulbs (ILB)" to encourage all parties concerned to stop using ILB by the end of 2013 for general lighting purposes and adopt more energy-efficient lighting.

As far as financial assistance is concerned, the HKSAR government launched a fund-supporting program, Building Energy Efficiency Fund Scheme (BEEFS), in 2009 which was set up to fund two types of environmental improvement projects, namely Energy-cum-carbon Audit Projects (ECA) and Energy Efficiency Projects (EEP). The applicants must be the building owners, owners' or residents' corporations. The total funding amount was HK$450 M, but the application was closed in April 2012. About 6400 buildings (about one-seventh of all buildings in HK) participated in and benefited from this program.

From the policies and regulations initiated by the HKSAR government, it is noted that all subsidies go to the owners of commercial buildings, not the tenants who are the actual decision makers of energy efficiency strategies and investment. Besides, there is a challenge of split incentives, especially for the case that the tenants are charged with an overall bill including management fee and electricity charges. In this case, limited information, such as how much electricity is saved, and saving benefits are provided to the tenants.

Institutional Perspectives on EETs Adoption

EETs adoption not only helps to reduce energy cost, but also to improve productivity, reliability, and working conditions (Worrell, Laitner, Ruth, & Finman, 2003). However, the productivity performance improvements are not the only driver of EETs adoption. Prior studies show that the widespread diffusion of EETs adoption is due to coercive, normative, or mimetic isomorphism (Moon & deLeon, 2007). Scholars agree that government and environmental groups are the main initiators that have shaped the institutional environment for EETs adoption (Bunse, Vodicka, Schönsleben, Brülhart, & Ernst, 2011; Kounetas & Tsekouras, 2008). For example, both commercial and industrial sectors are facing inducement-based coercion from US Government financial incentives to adopt EETs. Another example in HK retail context is the regulation from the HK Stock Exchange (HKSE). For listed companies in HK, HKSE requires all listed companies to comply with the ESG requirement (i.e., Appendix 27 Environmental, Social and Governance Reporting Guide) as requested by HK Exchange and Clearing Ltd. (which is the governing body of the HKSE) for the financial reporting period starting January 1, 2017. The ESG requirement asks for all HK listed companies to conform to all Key Performance Indicators (KPIs) which comprise Environmental Emissions and Use of Resources pertinent to the disclosure of related measures, initiatives, and results achieved.

Normative pressures from professional networks, academic institutions, and industry association networks do influence companies to adopt better EETs. In 2011, the launch of ISO 50001 (energy management system) asserts normative pressures on firms to adopt the energy management standard. In addition, EETs are increasingly popular among the competitors in their sector, causing additional mimetic pressure on the non-adopters. For example, when Walmart implemented energy-efficient retails in 2008, other retailers such as Kohl's Department Stores and Target appear to follow the same practices.

Table 9.1 Institutional perspectives on EET adoptions

Institutional perspectives		Effects
Normative pressure		Social norms arising from professional networks, academic institutions, and industry association to adopt energy management standard
Coercion pressure	Inducement-based	Providing incentives to organizations to conform to specific measures to EETs adoption
	Imposition-based	Using governmental regulations or industrial standards to demand energy-saving measures
Mimetic pressure		Mimetic pressure on non-adopters of EETs resulting in losing market competitiveness

(Scott, 1987)

From an institutional perspective, organizations not only compete for resources and customers, but also compete for the political power and institutional legitimacy. Firms seek to gain legitimacy in their environments in order to be accepted and thus ensure their long-term survival (Meyer & Rowan, 1977). Otherwise, they will be negatively sanctioned because of lack of perceived legitimacy (DiMaggio & Powell, 1983). Institutional theorists point out that pressures toward conformity can often lead to inefficient organization actions and structures sometimes (DiMaggio & Powell, 1983). Thus, firms take a strategic response on EETs adoption to obtain the institutional legitimacy, which may not receive better profitability in return.

Some scholars use institutional theory to explain how the institutional pressures impact on the diffusion of EETs adoption (Moon & deLeon, 2007; Zhu & Geng, 2010). However, which type of EEPs (imposition- and inducement-based coercion) is more effective to promote EETs remains unresolved. It is important to compare and contrast the impact of the two approaches that are commonly used by policy makers.

Two types of coercive power are discussed in the literature: imposition-based and inducement-based coercion in the institutional theory framework (Scott, 1987). While the former refers to mandates such as behaviors by law or industrial standards, the latter is a strategy that provides incentives to organizations to conform to specific conditions willingly (Scott, 1987). In this study, we define government financial incentives as inducement-based coercion, whereas regulatory policies as imposition-base coercion.

Table 9.1 summarizes the effects due to institutional perspectives.

Energy Sustainability in Luxury Retail Industry

Among all retail industries, it is noted that in order to highlight their displays and products, the luxury retail industry subscribes for a bigger and brighter outdoor lighting signpost for advertisement and attractiveness, and occupies more spacious areas resulting in using more air-conditioning and lighting electricity.

Limited research studies have been carried out to study how the retail industry deploys their energy sustainability strategy. Nevertheless, research results showed that cost savings, although not significant as compared to the overall cost expenses, and corporate social responsibility (Galvez-Martos, Styles, & Schoenberger, 2013; Richman & Simpson, 2016) are the two key factors contributing to the retail industry willingness to implement energy-saving strategies. Richman and Simpson (2016) compared two stores, of which one was an original store and the other a new replacement store. They found that deploying energy-reducing strategies in lighting and alternative space heating design at the new store gave an energy saving of 39 and 61%, respectively, as compared with the old store. Galvez-Martos, Styles, and Schoenberger (2013) looked into the best environmental management practices of retail trade sector in Europe and identified similar energy cost-saving initiatives. Apart from the bottom-line performance improvement which may not be significant regarding actual sum and pay-back periods, the corporate social responsibility is another momentum to organizations in pursuit of energy-saving implementations. Therefore, the aim of this study was to study the drivers and barriers of the luxury retail industry in EETs adoption.

Methodology

We employed a qualitative approach to understand the drivers and barriers of luxury retailers to adopt energy efficiency technologies in HK. The research is done through multiple case studies of personal interviews, company records, and published data sources (Yin, 1981) to reveal multivariate dimensions of the drivers and barriers being identified. Top managers of six luxury retail stores in HK were interviewed (see characteristics of the study participants in Table 9.2). The mix of the selected companies includes three different business sectors of luxury products, different nature, and size. In addition, to be considered as luxury product/brands, there are five dimensions, namely the perceived premium quality, the aesthetics, the expensiveness, the history, and the perceived utility of a luxury brand (Ghosh & Varshney, 2013). The selected participating companies are all fulfilling these five dimensions. The interview questions were sent to the interviewees before the interview which lasted for about one hour each.

As the interviewees received the questions well before the interviews, this allowed the interviewees to check with their companies on the content of their responses, resulting in more definite answers. The questions being asked were all open-ended in nature and related to their company policies on establishment of environmental policy, adoption of energy efficiency initiatives and tools, and publicity of energy-saving achievement. In the course of the interview, the drivers, barriers, and challenges that they were facing were also brought into the discussion. The interviews were then transcribed and analyzed within and across case(s). For

Table 9.2 Characteristics of the participants

Company	Business sector	Number of retail stores in HK (as of October 2016)	Time of interviews	Role of interviewee
A (publicly listed in HK)	Jewelry	40+	October 2016	Chief financial officer
B (publicly listed in HK)	Jewelry	70+	October 2016	Administrative manager
C	Watch	<10	October 2016	Managing director
D	Watch	<10	November 2016	Managing director
E	Fashion	40+	November 2016	Chief information officer
F (publicly listed in the USA)	Fashion	10+	December 2016	Senior director—information technology

the purpose of data triangulation, the results unveiled in the interviews were being compared with other sources of information, such as annual reports, company Web site information, and published information. All interviewees were guaranteed with anonymity, and their comments would not be traceable to any particular individual nor company.

Results and Analysis

Environmental and Energy Policy

Organizations are concerned with environmental sustainability so that the interview companies have developed their environmental policy, but surprisingly not an energy policy. Only the two participants responded that they would implement the energy policy in compliance with the ESG requirements of HKSE. Besides, all interview companies have implemented the energy saving in terms of usage management and design management on a company-wide basis.

[Respondents' comments]:

> Company A: "Being a listed company at HKSE, we would have to follow the ESG requirement starting January 2017 and shall report our energy saving achievement regarding Key Performance Indicators defined by us. We foresee this would be a difficult task as we did not have any experience in preparing such report but would abide by it."

> Company E: "We encourage colleagues to switch off the lights, air-conditioning, and any unnecessary devices when not in use, to keep the room temperature about 25 °C, to reduce printing emails or reports unless absolutely necessary, and to print both sides of papers."

Energy Cost Saving

In HK, it is well known that the property and labor costs are most expensive among the world. It is revealed that the electricity bill of all interview companies was only a small fraction ranging from 1 to 3% of their total expenses. They did not consider energy cost saving as the most important factor for their adoption of energy efficiency technologies or their strategy of energy saving. Rather, they considered it in terms of the corporate social responsibility orientation.

[Respondents' comments]:

> Company B: "We did not consider cost saving as one of our main energy efficiency factors, as the electricity cost is about 1% of our total expenses. We viewed our energy initiative as one of missions to show that we care for the environment."
>
> Company D: "Our electricity bill is very small compared to rental and salary expenses. We strive to be an environment conscious company to the eyes of our customers."

Adoption of EETs

All interview companies have replaced incandescent lightbulbs with LED lightbulbs for all display lightings. The relatively less sophisticated technologies, such as smart electricity meters, have not yet been employed by the companies, but only in a process of evaluation. Three of the interview companies were not even aware of smart electricity meters in the market. For renewable energy, it is hard for the retailers in HK due to the crowded environment. However, all interview companies indicated that they would be more than happy to try out any possible technologies or equipment to improve their energy-saving performance. For instance, one company was considering changing all air conditioners to those with energy efficiency labels, MEELS.

[Respondents' comments]:

> Company A: "We have changed most of our lightings to LED bulbs, and have started to evaluate the possibility of installing the smart electricity meters, in one part, to see if any possible saving be observed and, in other, to measure our performance with respect to the ESG requirement."
>
> Company B: "We did not receive much information regarding the latest technological development of energy efficiency tools. It will be beneficial to us if the government or related departments/institutions can help disseminate such information to us on a regular basis. We are keen on adopting any energy efficiency technologies as long as they are proved to work."
>
> Company C: "In addition to using LED light bulbs, we are also evaluating the possibility of replacing all our old air-conditioners with those with MEELS labels. One of the major concerns is related to the total equipment and installation costs and the time spent on replacement. The long replacement time will cause disturbance to our business. We hope the government can help subsidize the investment."

Supports and Incentives of Government Policies

We asked whether the interview companies noticed any current energy policies or support schemes from the HKSAR government that could help improve their energy consumption. Surprisingly, five interview companies were not aware of any such policies, except one company knew about the energy efficiency label scheme, MEELS, and two companies participated in the "Charter on External Lighting" advocated by the Environmental Bureau of the HK Government. Despite this lack of information, they suggested that financial supports in terms of tax rebate and/or investment subsidy on EETs adoption will make positive impact to the retail industry on reducing energy consumption and improving energy efficiency. The financial support should not merely direct to the building owners or landlords, but to the tenants as well. All companies even suggest that the government should provide alternative sources, such as solar energy, of clean energy for electricity.

[Respondents' comments]:

> Company B: "We are not aware of any government supports of energy efficiency plans, but we are welcome to such supports, particularly, in the form of financial subsidy. This will help us make our investment decision much easier and faster."

> Company C: "We know some government financial support which only goes to the building owners or landlords, not to us as a tenant. Government should devise a supporting scheme to benefit either both parties or the ultimate electricity users."

Drivers of Implementing Energy Sustainability Strategy

Inducement. There were four key inducement factors emerged from the study data: (1) corporate social responsibility, (2) top management's commitment, (3) corporation mission as a global company, and (4) responses to external incentives.

Corporate Social Responsibility

Organizations are all concerned to demonstrate their mission of being corporate social responsibility companies and their awareness and contribution in environmental and eco-friendly issues. Implementation of corporate social responsibility is getting more prominent especially for retail industry to enhance customer loyalty (Bolton & Matilla, 2015).

[Respondents' comments]:

> Company A: "We are a listed company and have to comply with the ESG requirement as required by the HKSE. Besides, we are environmentally conscious and will actively contribute to the corporate social responsibility as we consider this helps retain our customers."

Company B: "Our Chairman is very concerned with environmental sustainability issues. We even had one of our published publication dedicated to the topic of environmental protection. We want to show our customers that we are an environmentally friendly organization."

Top Management's Commitment—Mimetic Isomorphism

Energy sustainability strategy must be driven from the top-down approach as this involves commitment and resources. Top management also has more opportunities to mix with peer retail industry operators through social gatherings and public events. They can, therefore, mimic other retailers' best practices of energy sustainability strategy, as they might worry being the last one to follow suit resulting in losing customer loyalty and hurting corporate image both internally and externally.

[Respondents' comments]:

Company B: "Our top management would always bring back some energy saving ideas that he heard from the peer organizations. We also actively pursue to look for any energy efficiency technologies from Hong Kong Productivity Council or service providers."

Company C: "We always checked with our peer organizations through our gatherings organized by industry association any latest energy efficient development and tools. We also learnt or benchmark our energy efficient achievement with peer organizations and always sought for best practices in this area."

Corporate Mission as a Global Company

As most HK's luxury retail stores are multinational in nature invested by overseas global brand companies, a lot of energy sustainability strategies are stemmed from corporate mission as global companies are more forerunning in environmental policies and issues. Notably, the store design and use of energy efficiency tools are designated by the parent companies, i.e., the headquarters of those global companies, except overrode by local governmental rules and regulations.

[Respondent's comments]:

Company B: "As we have retail stores in Asia, we adopt the same retail layout design as far as energy efficiency is concerned. We would like to let our customers feel the same feeling whenever they shop with us in Asia."

Company F: "Most of our energy efficient measures and requirements are developed by our overseas headquarters. For the sake of corporate image, we will then follow suit. If there is any contradiction between our headquarters and local government, it is our policy to prevail the rules and regulations of local government."

Responses to External Incentives

This factor has two dimensions since HK retail operators are normally not the premises owners. The first dimension is whom the incentives be directed to and the second dimension is the substance of the incentives. As retail operators are the actual energy users who are responsible for the implementation of energy sustainability strategy, they should be the recipient of the incentives. The incentives from the local government should include but not limited to money (investment) rebate or tax incentive. It costs both money and time to replace any energy efficiency devices. For instance, chiller replacement may take weeks, if not months.

[Respondents' comments]:

> Company C: "We hope the government can provide subsidies on buying the new energy efficiency devices, especially the air-conditioners, of which these not only cost much but incur a large installation replacement fee as well."

> Company D: "We are not aware of any government incentives for energy saving or energy efficiency measures. Although the financial saving is not a large sum, any government incentives, in terms of tax rebate and investment subsidies, will be beneficial to us or SMEs at large."

Imposition. Two imposition factors emerged from the study data: (1) regulatory policies and (2) social norms.

Regulatory Bodies Policies—Coercive Isomorphism

In HK, all listed organizations of HK Stock Exchange (HKSE) are required to follow the ESG requirement (Appendix 27 Environmental, Social and Governance Reporting Guide) as requested by the HK Exchange and Clearing Ltd. (which is the governing body of the HKSE) for the financial reporting period starting January 1, 2017. This is not applied to private organizations, nor are there any rules and regulations being imposed by the HK government. All listed organizations are left without choice to conform to this ESG requirement which will definitely incur additional resources.

[Respondent's comments]:

> Company A: "We have devised our internal policy to follow the ESG requirement and implemented procedures to meet the requirements which are resource-demanding and need to prepare well ahead."

> Company B: "Being a listed company, we have to follow the ESG requirement starting January 2017."

Social Norm—Normative Isomorphism

Peer pressure from trade and industry associations will become a social norm for organizations to implement energy sustainability strategies. Through trade and industry associations, experiences and technology knowledge of deploying energy-saving practices and energy efficiency devices will be shared among organizations. This serves to exert pressure for those organizations not currently get up to par with others.

[Respondents' comments]:

> Company B: "We participate in the industry association and will obtain information from other companies in the mission of energy saving. We share knowledge and experience in these areas."
>
> Company F: "If we know any company using certain EETs with good results, we will sure evaluate such EETs to see if we can use them too."

Barriers of Implementing EETs

External barriers. Two factors identified as external barriers: (1) lack of control over the building and (2) competitive market conditions.

Lack of Control Over the Building

In HK, most retail stores are not owned, nor their building characteristics controlled, by the store operators. This limits the actual users in modifying the building envelope pertinent to energy sustainability consideration.

[Respondents' comments]:

> Company A: "Most of our shops are situated in the shopping mall or in the street, not owned by us. It is difficult for us to carry out any retrofits to save our electricity bill."
>
> Company D: "If we were to carry out any retrofits in the rented space, we have to re-condition it back to original shape and form which may not be justified as far as the energy saving is concerned."

Competitive Market Conditions

The market competitiveness intensity of the luxury retail industry is so fierce that the retailers are left without choice but to turn on their indoor lighting and exaggerate on their outdoor advertisement signpost in order to attract customers.

[Respondent's comments]:

Company A: "We need to stay being noticed by our customers and will keep our lightings on at some shops even after shopping hours."

Company C: "We are most concerned about our outdoor advertisement sign post of which our main focus is to attract our potential customers. While we have carried out studies and research to understand the color contrast at the sign post for brightness and attractiveness, we strive to use less distractive lightings to reduce light pollution at large."

Internal barriers. Three factors emerged as internal barriers: (1) relatively low cost savings from EETs, (2) lack of awareness of new technologies, and (3) lack of resources.

Low Cost Savings from EETs

Energy cost savings are relatively low with long pay-back times on the actual spending and investment in energy efficiency technologies and design although the saving achievement is large in a percentage term. This will in effect reduce to attract store owners to deploy energy sustainability strategies.

[Respondents' comments]:

Company B: "Although the saving in electricity cost is minimal compared to other costs, we strive for environmental protection in line with the advocate of the governmental policies and the retail industry."

Company E: "The electricity bill is relative small compared to other expenses, such as rental and staff costs. Our investment on energy efficiency technologies is not related to cost but to corporate social responsibility or customer loyalty instead."

Lack of Awareness of New Technologies

The retail industry is lack of information or awareness of latest energy efficiency technologies. This formation barrier is one of the major obstacle limiting organizations to adopt energy efficiency technologies (Kounetas, Skuras, & Tsekouras, 2011). This becomes a negative feedback as less retail stores are willing to invest that leads to less energy efficiency technology development in tandem with fewer suppliers to promote and sell energy-efficient devices.

[Respondents' comments]:

Company A: "We are not sure if the EETs work as said. We have to pilot test it by ourselves. This takes us a long time before we can make up our mind. We will be pleased if someone can provide us with some test results."

Company B: "We do not have any sources of getting the latest efficient technologies except being approached by those EETs suppliers occasionally. It will be of great assistance to us if we can receive or obtain more information and knowledge on EETs."

Lack of Resources

Resources constraint is another major factor of which organizations not only require to invest financially in the energy efficiency technologies, but provide necessary skilled human capital in operation as well.

[Respondents' comments]:

Company C: "We have to be selective in making our investment choices. Investment in terms of hard core money on energy saving may not be on the top of our investment list. But we do not mind saving environment with using less paper and electricity."

Company D: "As we are a small retail chain brand, we have limited resources and technical people to support any sophisticated energy efficiency technologies. We will only follow the rules and regulations as stipulated by the HKSAR government."

From the interview results, the organizations' environmental and energy policies, their energy cost saving, and the drivers and barriers of implementing energy sustainability strategies as summarized in Table 9.3 are identified in this study. The result findings are therefore used to hypothesize a conceptual model of Energy Sustainability Strategy Model in the following session.

The Conceptual Model of Energy Sustainability Strategy

From the interview results of six companies, using an institutional perspective, we collected empirical evidence that the luxury retail industry has been advocating energy-saving strategy, yet full-range implementations of EETs are still long way to go. We proposed a conceptual model of energy sustainability model based on the analysis of the drivers and barriers of how the luxury retail industry to adopt the EETs to improve their environmental sustainability performance (see Fig. 9.1). In particular, we studied their EETs adoption using the institutional model with respect to their environmental and energy policies, their energy cost saving, and their views of the supports and incentives of government policies.

Table 9.3 Drivers and barriers of EETs adoption

Drivers	Inducement	• Corporate social responsibility • Top management initiatives • Corporate mission as a global company • Responses to external incentives
	Imposition	• Regulatory bodies' policies • Social norm
Barriers	External	• Lack of control over the building • Competitive market conditions
	Internal	• Relatively low cost savings from EETs • Lack of awareness of new technologies • Lack of resources

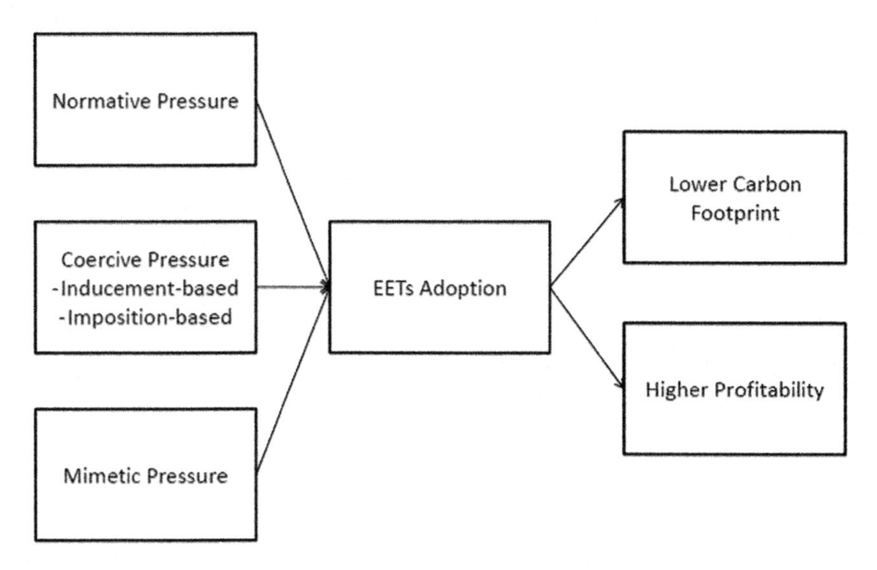

Fig. 9.1 A conceptual model of energy sustainability strategy

Figure 9.1 considers the normative, coercive, and mimetic pressures to delineate different energy sustainable strategies commonly deployed by organizations in recent years. Most organizations start off with considering how the energy be saved in the first place, as it is the easiest way to carry out. Subsequently, with the advancement of technologies and energy efficiency tools innovated with a better and yet affordable price performance ratio. The policy makers also play an important role in making organizations to consider energy-saving initiatives, organizations show much interests to use them. In short, organizations and policy makers can now have more options to advocate their energy sustainable strategies to not only yield a lower carbon footprint, but a higher profitability. Followings are focused to discuss EETs adoption using institutional perspectives.

Normative Pressure

Firms need to adopt EETs adoption by virtue of the pressures from professional institutions in order to stay competitive and sustain customer loyalty. Through the trade and associations, the retail industry is quick to learn from others in experience and technology knowledge of deploying energy-saving practices and energy efficiency devices. All interview companies have therefore developed their own environmental policies as like a standard among their profession. They considered this energy initiative as one of the missions to demonstrate their commitment to corporate social responsibilities. This is particularly important in luxury retail business as their customers are mostly environmental conscious and would like to buy from companies with similar mission in mind.

Besides, through the trade and industry associations, the retail industry cannot only share the experience and know-how of EETs adoption, but they can also seek for latest technologies and innovation in EETs through the cooperation with the education and research institutions as well. The industry can obtain financial supports from the HKSAR government in the research and development of EETs in collaboration with those institutions. The technology findings are to be shared among in the industry. This in effect provides a means of allowing the retail industry to share EETs adoption in practice. As such, the social norm of EETs adoption becomes common practices to the retail industry in implementing energy sustainability strategies.

Coercive Pressure

The inducement-based coercion is a strategy that provides incentives to firms to conform to certain requirements. Although the interview companies relate the actions of energy saving to their corporate social responsibilities and brand image, they expect the government can provide incentives to support their energy-saving policies and investment. It is hoped that the incentives will ultimately go directly to the energy-saving policy makers, that is the tenants in most commercial settings in HK, not just to the building owners or landlords. The incentives are expected in the form of investment subsidies and/or tax rebates. In addition, more publicity of government energy policies should be promoted to the organizations so as to provoke their active participation in energy saving.

The imposition-based coercion is another strategy that the lawmakers request certain behaviors by law or industrial standards. In HK, there are some measures as imposed by the government and institutional bodies. The HKSAR government has initiated a number of energy-related ordinances, such as Building Energy Efficiency Ordinance (BEEO), Building (Energy Efficiency) Regulations, and Energy Efficiency (Labeling of Products) Ordinance (EELPO), targeted to commercial and institutional buildings. Besides, the HK Stock Exchange requests listed organizations to comply with the ESG requirement (Environmental, Social, and Governance Reporting Guide) starting January 1, 2017. This ESG requirement has an enormous effect that requires the companies to put resources to meet this requirement.

Our results show that the imposition-based coercion has a larger effect than the inducement-based coercion, but we believe the latter will become an important factor if the tenants' benefits are also covered by the policy.

Mimetic Pressure

Firms copies certain successful management practices from other competing firms (Moon & deLeon, 2007). In EETs adoption, the same situations were found from the interview companies in particular for the energy saving in terms of usage management and design management.

For usage management, the organizations tend to use energy-efficient devices if the desired outcome can be achieved. For instance, they use lower power rating of lighting to avoid over-brightness. In addition, firms would install motion or infrared sensors to detect human presence to activate or deactivate electrical devices.

As far as the design management is concerned, firms include energy-saving strategies in their retail stores' layout design at the outset. In addition to utilize latest energy efficiency tools, the organization also considers the actual lighting and space management layout for product display and shopping comfort. A lot of technological advancements have progressed that make energy efficiency tools with improved performance at an affordable pricing. For instance, the LED lightbulbs are most widely found in lighting applications. Smart energy management device is also another commonly used tool for energy efficiency.

Organizations might have a good policy to pursue the corporate social responsibility strategy, but it is the frontline staff who actually operate and implement the energy-saving procedures. Company-wide awareness and a strong commitment to the corporate mission of energy saving shall explicitly be disseminated to all employees by the top management so as to receive rapports from them. Proper training of both energy-saving concepts and how to operate the energy efficiency tools should be duly carried out to equip all employees with necessary knowledge and know-how.

EETs Adoption

With regards to the institutional theory, firms will need to employ EETs adoption in order to stay competitive with their peers. From the research, it is found that all interview companies would try out or have implemented possible technologies to improve their energy-saving performance in meeting either management initiatives or requirements by local authorities. It is therefore expected that normative pressure, coercive pressure, and mimetic pressure will have direct and positive effects on EETs adoption.

Lower Carbon Footprint

With corporate social responsibility and environmental sustainability in mind, firms are willing to contribute to lower carbon footprint. By means of utilizing EETs adoption, such as latest energy-saving devices, firms will definitely help reduce the electricity usage which in essence lowers the carbon footprint. It is posited that EETs adoption will have direct and positive effects on reducing carbon footprint.

Higher Profitability

Although the interviewed firms suggest that the electricity cost saving is not significant with respect to other direct costs, such as rental cost and labor cost, this does contribute to the bottom line of the firm's profit. Besides, the EETs adoption leading to energy saving will demonstrate firm's corporate social responsibility to the public, which in turn will promote corporate branding and increase firm's business. As such, it is posited that EETs adoption will have direct and positive effects on high profitability.

Conclusions and Implications

The research investigates the EETs adoption of the luxury retail industry in HK. By using a qualitative analysis, six luxury retail companies are interviewed and studied. The results provide us with various viewpoints of their environmental and energy policies, energy-saving initiatives, EETs adoption strategy, and supports and incentives offered by the policy makers. From the results, we have identified drivers using institutional model and internal and external barriers of EETs adoption and also proposed an Energy Sustainability Strategy Model. This research advances more insights to understand the energy sustainability policies of the HK's luxury retail industry by and large.

The implications of this result have threefold. First, the organizations are faced with increasing demands of energy-saving measures in order to satisfy both their corporate mission and external stakeholders. The corporate social responsibility drives the organizations to take on the green initiatives, whereas the customers and policy makers request the implementation of energy efficiency measures. Secondly, the government policy makers should consider encouraging the organizations in EETs adoption using incentives, either investment subsidies or tax rebate. It is advisory that the incentives can go directly to the energy-saving policy makers. It is particularly important for those organizations who need to invest heavily in equipment replacement or upgrading. Thirdly, the organizations shall now begin to pursue energy sustainability strategies as this is starting to see more demand in the green initiatives being advocated by all stakeholders concerned.

Future studies on the HK's luxury retail industry should be in quantitative basis. A thorough analysis of how the proposed Energy Sustainability Strategy Model influences the luxury retail industry in EETs adoption should also be researched to enhance our understanding of their performance in energy saving. Finally, it is probable to provide the government policy makers with our findings so that they may consider putting up funding to encourage organizations in carrying out energy-saving initiatives, in particular the tenants in the retail sector.

References

Abdelaziz, E., Saidur, R., & Mekhilef, S. (2011). A review on energy saving strategies in industrial sector. *Renewable and Sustainable Energy Reviews, 15*(1), 150–168.

Bolton, L. E., & Mattila, A. S. (2015). How does corporate social responsibility affect consumer response to service failure in buyer–seller relationships? *Journal of Retailing, 91*(1), 140–153.

Bunse, K., Vodicka, M., Schönsleben, P., Brülhart, M., & Ernst, F. O. (2011). Integrating energy efficiency performance in production management—gap analysis between industrial needs and scientific literature. *Journal of Cleaner Production, 19*(6–7), 667–679.

DiMaggio, P. J., & Powell, W. W. (1983). The iron cage revisited: Institutional isomorphism and collective rationality in organizational fields. *American Sociological Review, 48*(2), 147–160.

Galvez-Martos, J., Styles, D., & Schoenberger, H. (2013). Identified best environmental management practices to improve the energy performance of the retail trade sector in Europe. *Energy Policy, 63*, 982–994.

Ghosh, A., & Varshney, S. (2013). Luxury goods consumption: A conceptual framework based on literature review. *South Asian Journal of Management, 20*(2), 146–159.

Kounetas, K., Skuras, D., & Tsekouras, K. (2011). Promoting energy efficiency policies over the information barrier. *Information Economics and Policy, 23*(1), 72–84.

Kounetas, K., & Tsekouras, K. (2008). The energy efficiency paradox revisited through a partial observability approach. *Energy Economics, 30*(5), 2517–2536.

Meyer, J. W., & Rowan, B. (1977). Institutionalized organizations: Formal structure as myth and ceremony. *American Journal of Sociology, 83*(2), 340–363.

Moon, S.G., & deLeon, P. (2007). Contexts and corporate voluntary environmental behaviors. *Organization & Environment, 20*(4), 480–496.

Pun, C. S., So, C. W., Leung, W. Y., & Wong, C. F. (2014). Contributions of artificial lighting sources on light pollution in Hong Kong measured through a night sky brightness monitoring network. *Journal of Quantitative Spectroscopy & Radiative Transfer, 139*, 90–108.

Richman, R., & Simpson, R. (2016). Towards quantifying energy saving strategies in big-box retail stores: A case study in Ontario (Canada). *Sustainable Cities and Society, 20*, 61–70.

Scott, W. R. (1987). The adolescence of institutional theory. *Administrative Science Quarterly, 32*(4), 493.

Worrell, E., Laitner, J.A., Ruth, M., & Finman, H. (2003). Productivity benefits of industrial energy efficiency measures. *Energy, 28*(11), 1081–1098.

Yin, R. K. (1981). The case study crisis: Some answers. *Administrative Science Quarterly, 26*(1), 58.

Zhu, Q., & Geng, Y. (2010). Drivers and barriers of extended supply chain practices for energy saving and emission reduction among Chinese manufacturers. *Journal of Cleaner Production, 40*, 6–12.

Chapter 10
Luxury Fashion and Peace Restoration for Artisans in Colombia

A Case Study in Luxury Fashion Designers and Artisans: Maestros Costureros "Master Sewers" Program

Cindy Cordoba

Abstract Sustainable luxury seeks for the ancestral meaning, quality of materials, and artisanship (Gardetti & Torres, 2003). According to the Alliance of Artisan Enterprise from the Aspen Institute, the artisan enterprise is valuable for native communities because it creates jobs and preserves ancient techniques (Aspen Institute, 2012). The program "Maestros Costureros" is an initiative that brings artisans and designers to collaborate, co-create, and discuss on design, artisanship, and luxury as a commitment to cultural heritage, peace, and economic development (Fucsia, 2016a). The artisans are from different regions in Colombia where the Colombian armed conflict existed for more than 50 years (Hough, 2011). The designers are scholars and technical experienced individuals in fashion design and management, with national and international exposure (Fucsia, 2016a). This initiative led by FUCSIA, a fashion magazine, and LCI Bogota with the support of Artesanías de Colombia, Inexmoda, and Club Colombia aims to create economic opportunities for artisans by facilitating their understanding of the current market through a creative collaboration with renowned fashion designers. The aim of this work is to explore the creative collaboration between artisan and designers through the analysis of "Master Sewers" program, the role of the different stakeholders, and how it relates to sustainability in the luxury fashion supply chain. The findings suggest that "Master Sewers" program is designed and implemented to improve the artisans' skill set in the design and marketing of their handmade products to fill the gaps in access to the marketplace. The fashion designers gained in the cooperation

The Master Sewers program is a co-creation program and multi-stakeholder initiative aims to revitalize the artisanal techniques and uplift the economic conditions of Colombia through the intervention of fashion designers with artisans for the production and commercialization of luxury fashion products.

C. Cordoba (✉)
Louisiana State University, Baton Rouge, LA, USA
e-mail: ccordo2@isu.edu

© Springer Nature Singapore Pte Ltd. 2018

C. K. Y. Lo and J. Ha-Brookshire (eds.), *Sustainability in Luxury Fashion Business*, Springer Series in Fashion Business, https://doi.org/10.1007/978-981-10-8878-0_10

by reaching a genuine source of inspiration that resonates with their roots. The products were exhibited in an official ceremony in LCI Bogota, an academic institution specialized in creative arts, and they are sold in a diversity of boutiques in Bogota, Colombia, South America (Artesanías de Colombia, 2016a). This research seeks to contribute to the Springer series in Sustainability in Luxury Fashion Business by providing a case study in the interconnection between artisanship and luxury for sustainability through the co-creation program "Master Sewers," a multi-stakeholder initiative for the revitalization of artisanal techniques and economic development for a better and more sustainable fashion.

Keywords Co-creation · Luxury · Sustainability · Artisanship
Social responsibility

Introduction

Luxury

Luxury refers to the creation of value that people perceived with a higher value compared to other products (Chevalier and Mazzalovo, 2008). Luxury products are used with a symbolistic aim of expressing well-being and status (Phau, Teah, & Chuah, 2015). They are appreciated by being rare and exclusive, and they have been associated with status-seeking and economic stratification (Kapferer, 2012). Luxury is more attributed to brands than to a concept (De Barnier et al., 2012), and these brands created value by generating a higher-level experience (Yeoman, 2011). The brand has a symbolic value combined with the prevalence given by the public to the designer in chief or creative director (Aakko, 2016, p. 38). This demonstrates the visibility and importance of designer in luxury fashion.

Sustainability in Luxury Fashion Business

Sustainable development is defined, as "development that meets the needs of the present without compromising the ability of future generations to meet their own needs" (Brundtland, 1987). Sustainability requires a responsible consumption of products (Brundtland, 1987). The inclusion of sustainability in fashion increases as consumer's demand from companies to be more sensitive and assume their social, economic, and environmental responsibility in their supply chain (Phau et al., 2015). Luxury fashion faces a paradoxical scenario (Black, 2008); on one hand, luxury promotes sustainability values by carefully crafted products, preservation of techniques, timeless and quality products (Kapferer & Michaut, 2015). By contrast, luxury has been associated with excess (Okonkwo, 2016) and serves a tool for stratification and inequality which is a disparity towards the concept of sustainable development (Hashmi, 2017). By addressing these issues, luxury brands should complement their excellence in quality by integrating an environmental and social purpose to both their business and communication strategy (Hennigs, Wiedmann, Klarmann, & Behrens, 2013).

According to several studies, consumers expect from luxury brands to be environmentally responsible (Kapferer & Michaut, 2015) for instance, by paying a higher price for acquiring a luxury product, consumers expect from luxury brands perform environmentally responsible practices (Kapferer & Michaut, 2015). Consumers tend to perceive highly crafted products and services that are considered luxuries as not being harmful to the environment (Cervellon, 2013). Among the attributes that are more prevalent to consumers in luxury is the quality of materials and durability which correlates with values promoted in sustainable consumption (Cimatti, Campana, & Carluccio, 2017). Other studies support the idea that when it is about luxury goods, consumer's purchase criteria prioritize product characteristics and ethical issues over environmental issues (Ki & Kim, 2016).

Luxury: The Slow Fashion Strategy

The common ground between artisanship and luxury is found in the promotion of long-lasting artisanship, high quality, and uniqueness (Black, 2008); values are transformed into sustainable benefits such as stronger human-product connections, long-lasting products, and reduction in overconsumption (Black, 2008); the concept of sustainable luxury and artisanship is connected through the bridge of slow fashion (Fletcher, 2008) which advocates for reducing production and consumption cycles through product quality and engagement (Jung & Jin, 2014). Slow fashion was born from the inspiration in the slow food movement created in Italy to promote the enjoyment of food against the arrival of fast food companies (Fletcher, 2008). The slow fashion movement started in Italy in 1986 (Cooper, 2005) and the term was coined by Angela Mumlis (Harris, Roby, & Dibb, 2016). Slow fashion advocates for the design and development of products of quality, unique pieces with a timeless character (Niinimäki and Hassi, 2011). It promotes fair labor, small lines, and durability (Pookulangara & Shephard, 2013). Slow fashion refers to an ethics of aware consumption, to a fully understanding of why and how things are produced and consumed (Fletcher, 2008). The reduction in production and consumption cycles is aimed to give the opportunity for the environment to recover from their extracted resources, and the need for achieving the balance between consumption and nature capacity is what guides slow fashion (Fletcher, 2008).

The relationship between artisanship and slow fashion can be explained by the fourth principles introduced by Jung and Jin (2014). Slow fashion advocates for the equity, which primarily refers to fair trade among the different stakeholders that integrate the production and consumption of the goods. On the second principle, slow fashion advocates for localism, which refers to the use of local resources to produce goods and the integration of artisanship. These characteristics lead us to the third principle that is authenticity; slow fashion products are loaded with meaning by the form and how they are developed. Exclusivity was also included as the fourth characteristic, which refers to the rarity that fashion items with uniqueness can deliver to the consumer, and functionality refers to the actual product use in context, which provides a sense of social status and identity (Niinimäki, 2010).

The decisions for the pursuit of slow fashion products are highly dependent on the consumer value systems (Niinimäki, 2010; Ozdamar & Atik, 2015). Zarley Watson and Yan (2013) emphasize that slow fashion consumers advocate for a long-term investment, quality, and product enjoyment, in contrast with fast fashion consumers who seek for low prices and trendiness (Black, 2008). Slow design not only seeks to fulfill the consumer needs with uniqueness but also seeks to fulfill the producer's life by providing a sense of accomplishment (Jung & Jin, 2014). For instance, companies create engagement between consumers and products by communicating stories associated with their products, who created them and how they made it (Fletcher, 2008). These stories increase the level of customer engagement and the willingness to pay for a premium product (Pookulangara & Shephard, 2013). By strengthening the level of engagement in the relation human—product, slow fashion seeks to reduce the fatigue for natural resource extraction (Pookulangar & Shephard, 2013) and it is an alternative for the creative burnout for designers produced by multiple several collections per year and the ready to wear apparel (Vincent, 2017). This seems to propose a solution to the production and consumption cycles and the environmental capacity to respond to both cycles (Fletcher, 2008).

Collaboration for Sustainable Development

Artisanship has been a source of exploration and innovation in design (Botnick & Raja, 2011). The exploration of indigenous artisanship provides a new source of creation for the designer, and it offers a return to a cultural heritage and a quest for uniqueness (Black, 2008); scholars have argued that innovation comes from the culture of artisanship, and it can provide a way to improve the quality of design (Bettiol & Micelli, 2014).

A wide range of preferences is created by a combination of social, economic, and cognitive characteristics (Spangenberg, 2013). In a globalized world, symbolism can survive in a mass production market due to the introduction of lifestyle groups (Lash & Urry, 1993). Since products are seen as symbolic resources (Thorpe, 2010), artisanal products have a place in a world where people are motivated to buy handicrafts for their artistic value (Dash, 2010). Artisanal products are produced completely by hand or the introduction of tools, as long as the majority of the development is manual (Tung, 2012). The artisanship is acquired by extensive practice and dissemination between the master and the apprentice in the case of rural artisans; the masters are family or local community members (Chuenrudeemol, Boonlaor, & Kongkanan, 2012).

Progressively, the competency increases between handicrafts and machine-made copies challenging artisans' visibility in the market and increasing the difficulties to trade their products (Dash, 2010). Among the difficulties for artisans to thrive in the market include limited infrastructure, distribution, financial accessibility, and finding an adequate market (Vadakepat, 2013). In the marketing channel side, artisans face other challenges including the cost of attending fair trades, competing with machine-made copies that claim to be handmade products, and promotion invisible crafts markets (Woolley & Sabiescu, 2015). These vulnerabilities in marketing strategies for artisans require the design of appropriate strategies for the special condition of handmade products (Vadakepat, 2013). The publication "Designers meet artisans" proposes the introduction of experienced designers to alleviate the artisan's difficulties, and the collaboration between designers and artisans plays a role in creating unique and quality products. The social engagement of design is illustrated by the designer's capacity to apply her/his skills to improve the living and economic conditions of people in need (Villari & Mortati, 2014). A growing awareness of social and environmental responsibility has led designers to work in collaboration with artisans by incorporating their techniques into their creations (Emmett, 2014). In the case of artisans, the distance between the lack of knowledge in the market and the artisanship is mediated by the collaboration between artisans and designers (Sethi, Duque, & Vencatachellum, 2005).

The above challenges faced by the artisans explain why the intervention of the designers and their social role in facilitating artisans' understanding of the market represents a sustainability outcome (Aakko, 2016). The collaboration between designer and artisan contributes to the designers' experience in providing a genuine source of inspiration, a return to a culture heritage and a quest for uniqueness (Black, 2008) and reducing the limitations for artisans including the gap of knowledge in design and marketing for contemporary needs (Sethi et al., 2005).

Co-creation

Co-creation saw as creative collaboration, a form of cooperation to achieve a shared purpose while mediating temperaments and individual objectives (Miell & Littleton, 2004), a negotiation to create innovation and growth (Mortati and Villari, 2013). Co-creation is catalyzed by the connection between different stakeholders' willingness to contribute purposely (Villari & Mortati, 2014). In co-creation, design and artisans come together to assemble their product design, services innovation, and marketing relationships (Alves, Fernandes, & Raposo, 2016). During designer–artisan collaboration, designers play a pivotal role to control the core essence of the art form (Kolay, 2016) while collaborating designers and artisans balance to preserve the essence of the traditional technique (Sethi et al., 2005). In some sense, the combination between designer and artisan aims to fulfill two needs; on the one hand, artisans who are trying to be reliable in a mass market with a different timing, and from the designer's side, the need to explore the cultural heritage and artisanship as a new source of inspiration (Tung, 2012). The designer serves a mediator to alleviate the disconnection between the past and the present (Sethi et al., 2005). The fulfillment of these two needs is materialized in products that couple ancestral and contemporary design (Murray, 2010).

Co-creation as a form of collaboration requires the ability of the designer to engage in the artisans' technique. With the purpose to support the creative endeavor between artisans and designers, UNESCO in partnership with Craft Revival Trust and Artesanías de Colombia published "Designers meet artisans," a framework for collaboration between artisans and designers, this work addresses the scope, strategies, and boundaries of design interventions to traditional crafts and techniques. It includes pedagogical strategies for collaboration between artisans and designers including workshops and creative laboratories using descriptive case studies in Colombia and India (Sethi et al., 2005). The main objectives of the designer's interventions are to reduce the gap between artisans and the market, enforce the artisanal labor, and contribute to the economic development of rural villagers.

Kapur and Mittar (2014) explore the means of design intervention for craft revival in India. This study included a survey of artisans and designers to identify the social, economic, and environmental outcomes of the design intervention in the artisanal dyeing process. The findings indicate that designers can serve as a bridge for artisans to the market and facilitate a better understanding of consumers' needs by providing training and technical support. Tung (2012) explores how designers collaborate with artisans to bring them closer to the market, their strategies and stages for their creative endeavors. This work uses a case study in rush-weaving artisans in Yuan Li, Taiwan, to illustrate the incorporation of traditional artisan technique in product development with contemporary design. It describes the synergy given between the designer and artisan from the idea to the product development. Both studies emphasize the role of the designer to facilitate and improve the performance of the artisans toward the marketplace. Scholars had argued that innovation truly comes from the culture of artisanship, and it can provide a way to improve the quality and success of design (Bettiol & Micelli, 2014) while the economic value of the collaboration between artisans and designers is presented in their ability to produce unique pieces (Tung, 2012).

Method

This research uses a case study method to explore the interception between artisanship and luxury for Sustainability in Luxury Fashion Business by analyzing the creative collaboration between artisans and designers in the "Master Sewers" program. This program embraces the concept of luxury and slow fashion in two approaches. The program embraces slow fashion concept by promoting the development of unique pieces with skilled labor, and it advocates for quality more than quantity, prioritizing craftsmanship over mass production (Clark, 2008). Here, the pieces result from a thoughtful exercise between designer and artisan pursuing the revival of artisan techniques. In addition, a sense of purpose guides the designers' intervention as a social contribution to close the gap between artisan and the market.

As luxury refers to exclusivity, well-crafted products, uniqueness, and durability are what is pursuit by Master Sewers through a dedicated work to deliver artisan pieces with a careful selection of materials (Joy, Sherry, Venkatesh, Wang, & Chan, 2012). "Master Sewers" promotes the development of the pieces with a high level of detail emerging modern fashion labels with traditional craftsmanship. The savoir faire of the artisans passed generation trough generation aims to deliver excellence in the handmade products as luxury pieces (Amatulli, Costabile, De Angelis, & Guido, 2017). By producing unique handmade items that include artisan techniques, beadwork, molas, draperies, embroidery, and cloth on canvas. For example, the inclusion of embroidered into cotton and viscose blouse shown in Fig. 1. The collaboration between artisans and designers contributes to the luxury strategy by the level of differentiation that the inclusion of artisan techniques can bring to the

Fig. 1 Cotton and viscose blouse with embroidery (Artesanías de Colombia, 2016b) created by the collaboration between the artisans, María Guzmán y Adriana Gomez, Joyce Rodriguez and designers, Diego Guarnizo y María Luisa Ortiz from the fashion brand "SOY" (Artesanías de Colombia, 2016b). The luxury element is presented in the inclusion of the Molas, a traditional artisan technique in Colombia that speaks heritage and genuineness

final product and the background associated with the artisan techniques and the people who produced them (Amatulli et al., 2017).

This work constitutes a case study, and case studies are open to a diversity of sources such as documents, reports, interviews, news, artifacts, and multimedia materials (Yin, 1984). Case study research provides details in the description of processes and attitudes (Yin, 1984). The case study serves as a form of description of a phenomenon and behavior, and its findings can be used toward the development of explanatory models (Fidel, 1984). The data collection contained a literature review in luxury, sustainability; creative collaboration focused on designer interventions. The study included an analysis of a diversity of sources (Rowley, 2002) corporate documents, newspaper articles, videos complemented with information provided by managers, designers, and artisans of the program. This study examined the publications by Revista Fucsia, which is one of the leader agents of the project, and the Revista Fucsia has dedicated a section on its Web site to cover Master Sewers program. This work also included a review of videos that record several scenes of the development of the program where participants described their experiences (LCI Bogotá, 2016a, b). To validate the results, this study has conducted interviews with participants of the program, designers, and artisans as well as representatives of the organizing entities.

Results

Program Structure

The program "Master Sewers" is an initiative that brings artisans and designers to collaborate co-create, learn, and discuss design, slow fashion, luxury, and artisanship (Fucsia, 2016a). The focus group is composed of ten renowned fashion designers (five from Colombia and five from foreign countries), and seven artisans from different regions of Colombia (Cauca, Atlántico, Valle, Nariño, and Valledupar) (Fucsia, 2016a). The artisan techniques included in this project are beadwork with chakiras, molas,[1] draperies, embroidery, and cloth on canvas (Fucsia, 2016a). The artisans are war survivors; they come different regions in Colombia where the armed conflict has existed more than 50 years (Hough, 2011). The designers are experienced in fashion design and management, with national and international exposure (Fucsia, 2016a).

The program "Master sewers" is created by Fucsia Magazine, supported institutionally by Artesanías de Colombia, LCI, and Inexmoda and financed by Club Colombia. During a four-month period, artisans travel to Bogota for learning, exchanging knowledge, and co-creating with designers. The program starts with the

[1]Mola: traditional textile skill called mola using a unique appliqué technique where patterns are formed by the shape and color of the ground material (Puls 1988).

20-days certified program “Hacia el lujo artisanal” which is designed and implemented by LCI Bogota, a higher educational institute specialized in creative arts. The first phase concluded with a design proposal of luxury apparel products within slow fashion principles. In this section, both designers and artisans have a space to know each other personally, their design interest, and personal lives. The second phase, co-creation of luxury products where designers and artisans mix their expertise to incorporate the traditional artisanship into a contemporary design that evocates the cultural diversity of Colombia (Figs. 2, 3, and 4).

As a result of a thoughtful design exercise, we can characterize slow fashion and luxury embedded in the Master sewers by its pursuit of products with a purpose using the design for economic development for artisans in underserved communities.

Phase 1: Training Program

The certified program “Hacia el lujo artisanal,” (towards the artisanal luxury) was distributed in different sessions and developed from May to August 2016 (Fucsia, 2016c) The certified program is designed and implemented by LCI Bogota, a formal education institution for creative arts in Colombia. The certified program is twenty days in length; during this period, one fashion designer with more than twenty years of experience specialized in artisanship provides a series of workshops

Fig. 2 Silk taffeta blouse that incorporated beadwork with chakiras created by the collaboration between the artisans Consuelo Campos and Enrique Tanigama and fashion designer Manuela Álvarez from the fashion brand “MAZ” (Artesanías de Colombia, 2016, b). The silk historically considered as a luxury material (Ivan, Mukta, Sudeep and Burak, 2016) combined with the beadwork applique provides a sense of exquisite and craftsmanship

Fig. 3 Wool cloth blouse that incorporated beadwork with chakiras created by the collaboration between the artisans Nelson Tanigama and Marisol Torres and fashion designer Kika Vargas. (Artesanías de Colombia, 2016b). The flowers are elaborated using beads, and the pieces are incrusted by hand into the wool cloth blouse. The beadwork highlights the background motifs

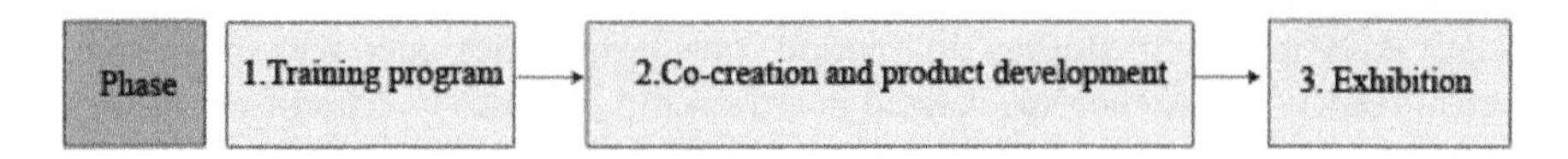

Fig. 4 "Master Sewers" program structure

in advanced techniques in pattern making and contemporary concepts in luxury design (Semana, 2016). This certified program is also a space to experiment with a diversity of materials by incorporating the traditional techniques that each of the seven artisans' domain. At the end of the program, the artisans graduated and LCI awarded them with the title of Master Sewers, a recognition that can serve a validation of high-quality products for participating in future trade events or commercialization opportunities (Fucsia, 2016b).

In terms of slow fashion and luxury, the training program aims to enhance the product development by integrating Haute couture techniques taught by mentors invited by LCI Bogota. Quality is a crucial aspect of the luxury strategy to guarantee the durability of the pieces. By applying these techniques opens an opportunity for the association of artisan products with expensiveness and quality (Fuchs, Schreier, & Van Osselaer, 2015).

Phase 2. Co-creation and Product Development

Artesanías de Colombia as the primary source of knowledge in crafts in Colombia allocates each artisan to a designer depending on designer's expertise and the artisan technique. Designers and artisans have a space to share their personal stories and design interests to create a friendly environment for the process of co-creation (LCI Bogotá, 2016a, b). This section continues with idea sketching by integrating the designer and artisan expertise and later prototyping the ideas resulting from a discussion between artisans and designers. In the product development, the prototyping of the products ideas tangible through experimentation (Villari & Mortati, 2014). Artisans and designers work together until the culmination of the co-created products.

In this phase, slow fashion is presented through the introduction of a thoughtful design exercise between designers and artisans, where they reflect the why and how things are produced and consumed (Fletcher, 2008), the preservation of tradition (Kapferer & Michaut-denizeau, 2014).

Phase 3. Exhibition

The program finished with an exhibition in LCI Bogota with the participation of designers, artisans, celebrities, influencers, bloggers, and representatives of the supportive organizations. After the exhibition of the products in the final ceremony, the products are displayed in boutiques and museum stores in Bogota, Colombia (LCI Bogotá, 2016a, b).

The exhibition enlarges the slow fashion experience because participants have the opportunity to appreciate the handmade work and explore the background of the pieces and their designers. The luxury argument is represented by a selected group of celebrities, influencers, and high-class network who endorsed the artisanal pieces promoted by the Fashion Magazine Fucsia, the level of exclusivity is the final touch to conclude the program and the exhibition is a demonstration to highlight the symbolic interaction between creator and costumer to sustain the hedonistic value of luxury products (Davies, Lee, & Ahonkhai, 2012).

Description of the Artisans

According to Artesanías de Colombia, the National Institute in Colombia for the study and promotion of Colombian crafts, around 58,821 people are dedicated to handicraft production, (Artesanías de Colombia, 2017). For native communities in Colombia, the artisanship is seen as a means of cultural value and economic development vehicle (Vega Torres, 2013). The artisan enterprise is of a high value for these communities because it creates jobs, preserves ancient techniques—and for the unique Colombian context—it promotes reconciliation, healing, and empowerment (Aspen Institute, 2012).

Colombian traditional artisanship includes natural fiber work, embroidery, molas, cloth on canvas, basketry, and beadwork. Native communities, cultural heritage, and other influences enrich this creative scholarship. In Colombia, the artisan enterprise represents the main source of income for more than

350,000 people, who represent the 15% of employment in the manufacturing industry (Luna & Rizo, 2011). Colombia presents a unique context, among the artisans responsible for artisanship, which are also located in the more affect regions of the armed conflict.

The seven artisans who are participants of "Master Sewers" come from several ethnic communities which are simultaneously the most affected by the Colombian armed conflict. The Colombian armed conflict initiated in the 1960s with the formation of the Revolutionary Armed Forces of Colombia—FARC (Hough, 2011) and then in the 80s their adversaries, the paramilitaries (Hristov, 2010), they had their origin in the employees of the private security of large landowners, industrialists, and regional politicians, drug trafficking, using this illegally armed structures for the acquisition of land (Hough, 2011). Paramilitary violence is responsible for most of the displacement in the country's conflict (Hristov, 2010). The armed conflict has been responsible for more than 220,000 (81.5% were civilians and 18.5%. Other crimes included massacres, enforced disappearances, torture, threats, kidnappings, anti-personal mines, economic (Centro Nacional de Memoria Histórica, 2013). A victim from the Colombian conflict has suffered one of the above crimes or has been affected by the suffering has been severely affected by the affectation to one relative member (Centro Nacional de Memoria Histórica, 2013). Participants of the Master Sewers program have suffered the consequences of the war and still have the resilience to contribute to the country development through their artisanship. The Transitional Justice refers to the term "victim" as any person who has suffered violations of human rights (Summers, 2012). In 2016, Colombia signed the Peace Agreement with acknowledging the suffering of the victims and demands to take measures to compensate and dignify the victims (Pearson, 2017). "Master Sewers" aims to address the post-conflict from the perspective of economic development for the victims and the role of the traditional artisanship as a form of art in peacebuilding (LCI Bogotá, 2016a, b).

Partners and Roles

This section identifies the several stakeholders that are part of the Master Sewers program; it describes their organizational nature and roles inside the program. Figure 5 summarizes the nature and roles of the partners of the program.

Leading Organization

Fashion Magazine Fucsia

The Fashion Magazine Fucsia is a woman's magazine specialized in fashion, culture, and lifestyle in Colombia. The magazine Fucsia was founded in 1999; it is a part of the journalism group "Semana" (Ramírez Verbel, 2013). The magazine is

Organizational nature	Private Sector	Public Sector	Public Sector	Private Sector	Private Sector	Private Sector
Partner's title	Fashion Magazine Fucsia	Colombian Agency for peace-building reintegration	Artesanias de Colombia	LCI -Bogota	Inexmoda	Club Colombia
Role in the "Master Sewers" Program	Coordinate the program activities and provide the platform for the development of the program and dissemination of the outcomes	Support the peace-building component of the program and the acknowledgment of the artisans who are simultaneously victims of the Colombian armed conflict.	Select the artisans considering need-based criteria and expertise in the artisan technique. It also provides provide scholastic support for the intervention artisans -designers.	Design and implement the curriculum of the certified program "Hacia el lujo artesanal" (Toward the artisanal luxury)	Provide scholastic support in the integration of artisanship into contemporary design	Sponsor the program activities, It serves as the financial anchor of the program as a part of the corporate social responsibility.
Role within slow fashion and luxury concepts						
Slow fashion	Connect a social purpose with a fashion design strategy, give a sense of purpose (Fletcher 2008) for the production of handmade products	Endorse program for improvement of socio-economic development of undeserved communities for economic (Blowfield 1999)	Help to reduce the gap between artisan the market by facilitating the technical support for production and commercialization of handmade products (Sethi, Duque, and Vencatachellum 2005). .	Provide a level of understanding of the why and how things are produced and experiences or consumed (Fletcher 2008)	Provide the scholar support to characterize the artisan techniques, as high value work for a luxury strategy	Endorse programs that associated brands with ethical and social pursuits (Lamrad and Hanlon 2014) and elevated traditional knowledge in Colombia craftsmanship.
Luxury	Connect a strong celebrities influencer and upper class network to endorse the luxury products resulting from the collaboration between designers and artisans.	Promotes ethics as a competitive advantage for companies to highlight their positive impact in ther communities and economic capability,. (Partridge 2011)	Preserve traditional artisan techniques, Offer protection to differentiate handmade products from mass production markets using labels (Artesanias de Colombia, 2016a)	Integrate a thoughtful selection of materials and techniques with contemporaneity design concepts, from idea sketching to product development.	Promote Master sewers products as luxury pieces, durable of limited edition (Niinimäki 2015).	Highlight the aesthetic value of artisan products based on their expertise in Colombian craftsmanship. (Seo and Buchanan-Oliver 2015)

Fig. 5 Partner's nature and role in the "Master Sewers" program

distributed biweekly, and readers can find the magazine online and in retail stores. Through the years, Fucsia Magazine has been established as a Colombian fashion authority and a guide for style and living (Ramírez Verbel, 2013). The role of the Fucsia Magazine in "Master Sewers" is to serve as the program's leading organization, which coordinates the activity, provides the platform for the development of the program and dissemination of the outcomes (Fucsia, 2016a). The fashion magazine also plays a key role by providing a network of fashion companies, influencers, and institutional representatives to embrace the Master Sewers program for social advocacy (Fucsia, 2016a).

Institutional Partners

Artesanías de Colombia

Artesanías de Colombia founded in 1964 as the governmental organization responsible for the promotion and development of the craft sector in Colombia (Sethi et al., 2005). Its role in "Master sewers" is to select the artisans to participate in "Master Sewers" from its official registered database of artisans by considering need-based criteria and technical expertise based on active participation in events promoted by Artesanías de Colombia.

Colombian Agency for peacebuilding reintegration

As a governmental organization for peacebuilding. Colombian Agency for peacebuilding reintegration—ACR works for the acknowledgment of the Colombian armed conflict victims, reintegration of members from the armed conflict, and restorative initiatives. In this perspective, ACR includes economic development as a crucial component for peacebuilding, which aims to dignify the victims beyond the compensation measures. The role of this organization in "Master Sewers" is to support the peacebuilding component of the program. (Agencia Colombiana para la Reintegración, 2011) and acknowledge their participants as war victims and the need to implement this initiative as a restorative measure.

Academic Partner

LCI Bogota

LCI is a higher educational network founded in 1959 in Canada, and this educational network has 22 campuses on five continents; LCI Colombia was established in 1997, and this institution provides creative careers including fashion design, jewelry, creative industries management, marketing, and communication in fashion (LCI Bogotá, 2016a, b). The role of this organization in the program was to design and implement the curriculum in the certified program "Towards the luxury fashion." LCI made use of its network of renowned fashion designers scholars to organize a curriculum that provides a contemporary view of the design for the artisans.

Corporate Partners

Inexmoda

Institute for Export and Fashion—INEXMODA in Colombia was founded in 1987 (Inexmoda, 2017). This organization is a private organization, and it is the primary source of information, research, marketing, and training for the fashion industry in Colombia. This institute created the fashion show Colombiamoda and the textile industry fair, Colombiatex, which are the main platforms for the exposure of the Colombian Fashion Industry (Malaver, Rivera, Sierra, & Cardona, 2014). Inexmoda considers the fashion industry in Colombia as a system; in that sense, Inexmoda works for providing support to increase competitiveness for the industry (Inexmoda, 2017). The role of Inexmoda in Maestros Costureros is to serve as a scholarship advisor in Colombian artisanship because they are one of the relevant organizations working for the development of the fashion industry in Colombia. Inexmoda has a Fashion and Economic Laboratory, where one of their approaches is to explore the Colombian artisanship to integrate with contemporary concepts of fashion, from this analysis; Inexmoda provides recommendations for clothing and textile companies in Colombia.

Club Colombia

Club Colombia is a largely premium beer company in Colombia, which advocates for the local wisdom and the integration of the Colombian tradition into beverages (Fucsia, 2016a). They sponsor the "Master Sewer" program as a corporate responsibility strategy to promote Colombia cultural heritage and ownership by advocating for traditional values that are represented in the Colombian artisan techniques for cultural ownership and promotion of artisans and designers (Fucsia, 2016b).

Conclusion

In emerging economies, there is a need for using design as a transformative tool for social and economic growth (Barker & Hall, 2009). "Master sewers" presents a synergy between different stakeholders in which each one plays a role from the position and expertise they know better; for the designers, their experience implies an understanding of the market contemporaneous needs and a strength to share with the artisans. For the artisans, their techniques learned through generations represent a valuable source of inspiration and mastery; Fucsia Magazine serves as an articulating axis for the implementation of the program, Artesanías de Colombia provides the scholastic support in Colombian artisan techniques, LCI brings the creative and technical understanding of the market, and Inexmoda contributes with their knowledge in the fashion industry and how artisanship is incorporated in sustainable fashion manner, and Club Colombia serves as an sponsor within their corporate social responsibility strategy to highlight artisanship and tradition. The alignment of these several stakeholders is the foundation for the development of the Master sewer program and the support to local communities (Chapple & Moon, 2005).

"Master sewers" seeks to revitalize the Colombian artisanship by the designer's intervention, and the collaboration between artisans is an invitation to explore ancestral techniques from a genuine source. The co-created products are the result of a conversation between the past and the present in the Colombian culture. The harmonization and contrasting colors visible in the design evokes diversity in techniques and design perspectives. This interaction not just contributed to the advance of the artisan, and it is also a renewal for the designer to face an uncommon source of inspiration and creativity liberation (Tung, 2012). The designer's intervention to the artisanship is an effort for understanding the artisans' native culture (Murray, 2010). The fashion designs in the artisanal luxury pieces expose an ancestral and ceremonial meaning of fabrics colors, elements, and exotic textures.

The product results of this creative collaboration are created to be accessible for people with an understanding of their artistic value and willingness to pay fair prices for handmade products. They are not produced in massive quantities due to they carry a time-consuming labor. The crucial aspect is to promote the luxury handmade products as pieces of art than as regular products (Kapferer, 2012). The challenges are to balance the timing in artisanship and the consumer's expectancies, where the designers play a role as mediator. The artisanship not only has a price at the economic level but also a great value that transcends through its authenticity (seen as a ritual).

The implication in terms of slow fashion concepts is the design practice of interconnection between designer and artisans' perspective to elaborate a crafted product that evocates the cultural heritage of the artisans simultaneously resonating with a contemporary market. In regard to luxury, the handmade pieces integrated wisely the traditional techniques built generation upon generation; the endorsed of fashion designers highlights the importance of the preservation of these artisan techniques and the impact of helping artisans to understand the market and how they can take advantage of their skilled work in the elaboration of handmade pieces to be priced as luxury pieces to contribute to better earning and in long-term to their economic development to improve their economic development.

Beyond the luxury fashion products, "Master Sewers" represents a peacebuilding strategy that elevates the cultural value of artisans who belong to ethnic groups highly affected by the Colombian conflict. Communities have faced for more than 50 years the pain of war left by the presence of armed groups in their territories. The arts change the discourse around conflict and peace for their ability to inspire, memorialize, and unmask hidden truths. Artisan enterprises as a natural form of art for indigenous, and Afro-Colombian communities have served as an economic force to uplift their living conditions because it generates income and elevates the cultural values that these communities have vigorously preserved over time. Integral restoration is defined as the series of measures aim to restitute the victims to the state in which the events of victimization occurred (Zernova, 2007). Within this context, the implementation of strategies restores the victims' enterprises, by recreating the environment for economic development and integral reparation. The sustainability outcome of the creative collaboration between artisans and designers relies on the designers' social role of improving the chances of the artisans to participate in the marketplace, revitalization of artisan techniques incorporated in contemporary design for luxury pieces, and the sense of achievement experienced by the artisans for having the opportunity to share their knowledge with fashion designers.

References

Aakko, M. (2016). Fashion in between, artisanal design and production of fashion. *Dissertation*. Espoo: Aalto University. http://urn.fi/URN:NBN:fi:aalto-201701201341. Accessed May 9, 2017.

Agencia Colombiana para la Reintegración (ACR). (2011). *Agency*. http://www.reintegracion.gov.co/en/agency/Pages/default.aspx. Accessed May 9, 2017.

Alves, H., Fernandes, C., & Raposo, M. (2016). Value co-creation: Concept and contexts of application and study. *Journal of Business Research, 69*(5), 1626–1633.

Amatulli, C., Costabile, M., De Angelis, M., & Guido, G. (2017). Inside luxury: Main features, evolving trends, and marketing paradoxes. In C. Amatulli, M. De Angelis, M. Costabile, & G. Guido (Eds.), *Sustainable luxury brands: Evidence from research and implications for managers* (pp. 7–34). UK, London: Palgrave Macmillan. https://doi.org/10.1057/978-1-137-60159-9_2.

Artesanías de Colombia. (2016). *Informe de gestion 2016*. http://www.artesaniasdecolombia.com.co:8080/PortalAC/C_sector/caracterizacion_81. Accessed May 9, 2017.

Artesanías de Colombia. (2016a). *"Maestros costureros" para compartir la tradición*. http://artesaniasdecolombia.com.co/PortalAC/Noticia/maestros-costureros-para-compartir-la-tradicion_8554. Accessed May 18, 2017.

Artesanías de Colombia. (2016b). *"Maestros costureros" photo exhibition*. Bogota.

Artesanías de Colombia. (2017). *Caracterización del sector artesano en Colombia* [Characterization of the artisan sector in Colombia]. http://www.artesaniasdecolombia.com.co:8080/PortalAC/C_sector/caracterizacion_81. Accessed May 9, 2017.

Aspen Institute. (2012). *Alliance for Artisan Enterprise, bringing Artisan Enterprise to Scale*. https://assets.aspeninstitute.org/content/uploads/files/content/images/Alliance%20for%20Artisan%20Enterprise%20Concept%20Note_0.pdf. Accessed May 07, 2016.

Barker, T., & Hall, A. (2009). Go global: How can contemporary design collaboration and e-commerce models grow the creative industries in developing countries? In Korean Society of Design Science (Ed.), 2009: *International association of societies of design research* (pp. 2227–2236).

Bettiol, M., & Micelli, S. (2014). The hidden side of design: The relevance of artisanship. *Design Issues, 30*(1), 7–18. https://doi.org/10.1162/DESI_a_00245.

Black, S. (2008). *Eco-chic: The fashion paradox*. UK: Black Dog Publishing.

Botnick, K., & Raja, I. (2011). Subtle technology: The design innovation of Indian artisanship. *Design Issues, 27*(4), 43–55.

Brundtland. (1987). *Report of the world commission on environment and development: our common future*. United Nations.

Centro Nacional de Memoria Histórica. (2013). *¡ Basta ya! Colombia: Memorias de guerra y dignidad* [No anymore! Colombia: Memories of war and dignity]. Bogota: Centro Nacional de Memoria Histórica.

Cervellon, M. C., & Shammas, L. (2013). The value of sustainable luxury in mature markets: A customer-based approach. *The Journal of Corporate Citizenship, 52,* 90.

Chapple, W., & Moon, J. (2005). Corporate social responsibility (CSR) in Asia: A seven country study of CSR web site reporting. *Business and Society, 44*(4), 415–441.

Chevalier, M., & Mazzalovo, G. (2008). *Luxury brand management: A world of privilege*.

Chuenrudeemol, W., Boonlaor, N., & Kongkanan, A. (2012). Design process in retrieving the local wisdom and communal identity: A case study of Bangchaocha's bamboo basketry crafts. In *Proceedings of the 6th International Conference of Design Research Society* (pp. 1–4).

Cimatti, B., Campana, G., & Carluccio, L. (2017). Eco design and sustainable manufacturing in fashion: A case study in the luxury personal accessories industry. *Procedia Manufacturing, 8,* 393–400.

Clark, H. (2008). SLOW + FASHION—an Oxymoron—or a Promise for the Future …? *Fashion Theory, 12,* 427–446. https://doi.org/10.2752/175174108X346922.

Cooper, T. (2005). Slower consumption reflections on product life spans and the "throwaway society". *Journal of Industrial Ecology, 9*(1–2), 51–67.

Dash, M. (2010). Buyers' preferences for purchase of selected handicrafts with special reference to Orissa. *IUP Journal of Management Research, 9*(6), 38.

Davies, I. A., Lee, Z., & Ahonkhai, I. (2012). Do consumers care about ethical-luxury? *Journal of Business Ethics, 106,* 37–51. https://doi.org/10.1007/s10551-011-1071-y.

De Barnier, V., Falcy, S., & Valette-Florence, P. (2012). Do consumers perceive three levels of luxury? A comparison of accessible, intermediate and inaccessible luxury brands. *Journal of Brand Management, 19*(7), 623–636.

Emmett, D. (2014). Conversations between a foreign designer and traditional textile artisans in India: Design collaborations from the artisan's perspective.

Fidel, R. (1984). The case study method: A case study. *Library and Information Science Research, 6*(3), 273–288.

Fletcher, K. (2008). *Sustainable fashion and textiles: Design journeys*. Sterling, London: Earthscan.

Fuchs, C., Schreier, M., & Van Osselaer, S. M. J. (2015). The handmade effect: What's love got to do with it? *Journal of Marketing, 79,* 98–110. https://doi.org/10.1509/jm.14.0018.

Fucsia. (2016a). *Comienza el proyecto maestros costureros* [The program "Master Sewers starts]. http://www.fucsia.co/especial/especiales-comerciales/articulo/comienza-el-proyecto-maestros-costureros-2016/71604. Accessed April 12, 2017.

Fucsia. (2016b). *Un taller de lujo y creation* [A workshop of luxury and creation]. http://www.fucsia.co/especial/especiales-comerciales/galeria/maestros-costureros-2016-taller-de-co-creacion/71771. Accessed May 24, 2017.

Fucsia. (2016c). *Hacia el lujo artisanal* [Toward the artisanal luxury]. http://www.fucsia.co/contenidos-editoriales/maestros-costureros-2016/articulo/artesania-colombiana-y-lujo-artesanal/73000. Accessed May 24, 2017.

Gardetti, M., & Torres, A. (2003). *Entrepreneurship, innovation and luxury* (pp. 55–75). UK: Greenleaf Publishing.

Harris, F., Roby, H., & Dibb, S. (2016). Sustainable clothing: Challenges, barriers, and interventions for encouraging more sustainable consumer behavior. *International Journal of Consumer Studies, 40*(3), 309–318.

Hashmi, G. (2017). Redefining the essence of sustainable luxury management: The slow value creation model. In *Sustainable management of luxury* (pp. 3–27). Singapore: Springer.

Hennigs, N., Wiedmann, K. P., Klarmann, C., & Behrens, S. (2013). Sustainability as part of the luxury essence: Delivering value through social and environmental excellence. *The Journal of Corporate Citizenship, 52,* 25.

Hough, P. A. (2011). Guerrilla insurgency as organized crime: Explaining the so-called "Political Involution" of the revolutionary armed forces of Colombia. *Politics & Society, 39*(3), 379–414. https://doi.org/10.1177/0032329211415505.

Hristov, J. (2010). Self-defense forces, warlods, or criminal gangs? Toward a new conceptualization of paramilitarism in Colombia. *Labour: Capital and Society,* 43–56.

Inexmoda. (2017). About http://www.inexmoda.org.co/Inexmoda/QuienesSomos/tabid/259/Default.aspx. Accessed February 15, 2017.

Ivan, C.-M., Mukta, R., Sudeep, C., Burak, C. (2016). Long-term sustainable sustainability in luxury. Where else? In M. A. Gardetti, & S. S. Muthu (Eds.), *Handbook of sustainable luxury textiles and fashion: Volume 2* (pp. 17–34). Singapore: Springer. https://doi.org/10.1007/978-981-287-742-0_2.

Joy, A., Sherry, J. F., Venkatesh, A., Wang, J., & Chan, R. (2012). Fast fashion, sustainability, and the ethical appeal of luxury brands. *Fashion Theory, 16,* 273–295. https://doi.org/10.2752/175174112X13340749707123.

Jung, S., & Jin, B. (2014). A theoretical investigation of slow fashion: Sustainable future of the apparel industry. *International Journal of Consumer Studies, 38*(5), 510–519.

Kapferer, J., & Michaut-denizeau, A. (2014). Is luxury compatible with sustainability? Luxury consumers' viewpoint. *Journal of Brand Management, 21*(1), 1–22. https://doi.org/10.1057/bm.2013.19.

Kapferer, J. N. (2012). Abundant rarity: The key to luxury growth. *Business Horizons, 55*(5), 453–462.

Kapferer, J. N., & Michaut, A. (2015). Luxury and sustainability: A common future? The match depends on how consumers define luxury. *Luxury Research Journal, 1*(1), 3–17.

Kapur, H., & Mittar, S. (2014). Design intervention & craft revival. *International Journal of Scientific and Research Publications, 4,* 10.

Ki, C. W., & Kim, Y. K. (2016). Sustainable versus conspicuous luxury fashion purchase: Applying self-determination theory. *Family and Consumer Sciences Research Journal, 44*(3), 309–323.

Kolay, S. (2016). Cultural heritage preservation of traditional Indian art through virtual new-media. *Procedia-Social and Behavioral Sciences, 225,* 309–320.

Lash, S., & Urry, J. (1993). *Economies of signs and space* (Vol. 26). Thousand Oaks: Sage Publications.

LCI Bogotá. (2016a). Maestros Costureros, es una muestra del compromiso por la moda y el país, video recording. *YouTube*. Viewed July 17, 2016. https://www.youtube.com/JGbUryfcbbw.

LCI Bogotá. (2016b). *Quienes somos* [About us]. http://www.lci.edu.co/conocenos/historia. Accessed July 9, 2017.

Luna, R., & Rizo, J. (2011). Caracterización socio-económica y desempeño productivo de las microempresas artesanales en la ciudad de Santa Marta (2000–2009). *Dimension Empresarial, 9*(1), 115–130.

Malaver, M. N., Rivera, H. A., Sierra, M., & Cardona, D. F. (2014). Art and strategy: The case study of the fashion industry in Colombia. *Pensamiento & Gestión, 36,* 184–205.

Miell, D., & Littleton, K. (2004). *Collaborative creativity: Contemporary perspectives*. Free Association Books.

Mortati, M., & Villari, B. (2013). Crafting social innovators: Designing collaborative, participative, networked solutions in urban contexts. *Craft + Design Enquiry, 5*, 125–140.

Murray, K. (2010). Outsourcing the hand: An analysis of craft-design collaborations across the global divide. *Craft + Design Enquiry, 2*, 1–23.

Niinimäki, K. (2010). Eco-clothing, consumer identity and ideology. *Sustainable Development, 18* (3), 150–162. https://doi.org/10.1002/sd.455.

Niinimäki, K., & Hassi, L. (2011). Emerging design strategies in sustainable production and consumption of textiles and clothing. *Journal of Cleaner Production, 19*(16), 1876–1883.

Okonkwo, U. (2016). *Luxury fashion branding: Trends, tactics, and techniques*. UK: Palgrave Macmillan.

Ozdamar, Z., & Atik, D. (2015). Sustainable markets: Motivating factors, barriers, and remedies for mobilization of slow fashion. *Journal of Macromarketing, 35*(1), 53–69.

Pearson, A. (2017). Is restorative justice a piece of the Colombian transitional justice puzzle? *Restorative Justice, 5*(2), 293–308.

Phau, I., Teah, M., & Chuah, J. (2015). Consumer attitudes towards luxury fashion apparel made in sweatshops. *Journal of Fashion Marketing and Management, 19*(2), 169–187.

Pookulangara, S., & Shephard, A. (2013). Slow fashion movement: Understanding consumer perceptions—An exploratory study. *Journal of Retailing and Consumer Services, 20,* 200–206. https://doi.org/10.1016/j.jretconser.2012.12.002.

Puls, H. (1988). *Textiles of the Kuna Indians of Panama*. London: Shire Publications Ltd.

Ramírez Verbel, C. (2013). *El periodismo está de moda una aproximación al oficio del periodismo de moda en Colombia* [Jism is in fashion an approach to the office of fashion Jism in Colombia] (Bachelor thesis). Pontifica Universidad Javeriana, Colombia. https://repository.javeriana.edu.co/bitstream/handle/10554/14653/RamirezVerbelCarolina2013.pdf?sequence=1. Accessed May 18, 2017.

Rowley, J. (2002). Using case studies in research. *Management Research News, 25*(1), 16–27.

Semana. (2016). *Llega el diplomado que homenajea a los "intérpretes" de la moda* [The certified program that honors the "interpreters" of fashion]. http://www.semana.com/educacion/articulo/diplomado-de-moda-tecnicas-y-conceptos-contemporaneos-del-lujo-artesanal-en-colombia/475273. Accessed May 18, 2017.

Sethi, R., Duque, C.D., & Vencatachellum, I. (2005). *Designers meet artisans: A practical guide*. Craft Revival Trust, Artesanías de Colombia & UNESCO.

Spangenberg, J. H. (2013). Design for sustainability (DfS): Interface of sustainable production and consumption. *Handbook of sustainable engineering* (pp. 575–595). The Netherlands: Springer.

Summers, N. (2012). Colombia's victims' law: Transitional justice in a time of violent conflict? *Harvard Human Rights Journal, 25*(1), 219–235.

Thorpe, A. (2010). Design's role in sustainable consumption. *Design Issues, 26*(2), 3–16.
Tung, F. W. (2012). Weaving with rush: Exploring craft-design collaborations in revitalizing a local craft. *International Journal of Design, 6*(3), 71–84. http://www.ijdesign.org/ojs/index.php/IJDesign/article/viewFile/1077/528. Accessed May 9, 2017.
Vadakepat, V. M. (2013). Rural retailing: Challenges to traditional handicrafts. *Journal of Global Marketing, 26*(5), 273–283.
Vega Torres, D. R. (2013). El Aprendizaje de la Artesanía y su Reproducción Social en Colombia [The learning of handicrafts and their social reproduction in Colombia]. *Educación y Territorio, 2*(1), 89–112.
Villari, B., & Mortati, M. (2014). Design for social innovation: Building a framework of connection between design and social innovation. In *Proceedings of the fourth Service Design and Service Innovation Conference*, Lancaster University, United Kingdom, 9–11 April 2014 (No. 099, 79–88). Linköping: University Electronic Press.
Vincent, A. (2017). Breaking the cycle: How slow fashion can inspire sustainable collection development. *Art Libraries Journal, 42*(1), 7–12. https://doi.org/10.1017/alj.2016.42.
Woolley, M., & Sabiescu, A. (2015). The use of craft skills in new contexts. http://www.digitalmeetsculture.net/wp-content/uploads/2015/09/RICHES-D5.1-The-Use-of-Craft-Skills-in-New-Contexts_public.pdf. Accessed April 18, 2017.
Yeoman, I. (2011). The changing behaviors of luxury consumption. *Journal of Revenue and Pricing Management, 10*(1), 47–50.
Yin, R. K. (1984). *Case study research*. Thousand Oaks: Sage Publications.
Zarley, M., & Yan, N. (2013). An exploratory study of the decision processes of fast versus slow fashion consumers. *Journal of Fashion Marketing and Management: An International Journal, 17*(2), 141–159.
Zernova, M. (2007). *Restorative justice: ideals and realities*. Hampshire: Ashgate.

Index

© Springer Nature Singapore Pte Ltd. 2018

C. K. Y. Lo and J. Ha-Brookshire (eds.), *Sustainability in Luxury Fashion Business*,
Springer Series in Fashion Business, https://doi.org/10.1007/978-981-10-8878-0

S

T

W

Z

Printed by Printforce, the Netherlands